Social Perspectives on the Sanitation Challenge

Bas van Vliet · Gert Spaargaren
Peter Oosterveer
Editors

Social Perspectives on the Sanitation Challenge

Springer

Editors
Bas van Vliet
Environmental Policy Group
Wageningen University
Hollandseweg 1
6706 KN Wageningen
The Netherlands
Bas.vanVliet@wur.nl

Gert Spaargaren
Environmental Policy Group
Wageningen University
Hollandseweg 1
6706 KN Wageningen
The Netherlands
Gert.Spaargaren@wur.nl

Peter Oosterveer
Environmental Policy Group
Wageningen University
Hollandseweg 1
6706 KN Wageningen
The Netherlands
Peter.Oosterveer@wur.nl

ISBN 978-90-481-3720-6 e-ISBN 978-90-481-3721-3
DOI 10.1007/978-90-481-3721-3
Springer Dordrecht Heidelberg London New York

Library of Congress Control Number: 2010922051

Printed on acid-free paper

Springer is part of Springer Science+Business Media (www.springer.com)

Preface

This volume is the result of years of commitment with world-wide sanitation challenges from various research networks linking the editors and authors of this volume to many other sanitation scholars and professionals. Major contributions to this volume are derived from the work done in the PROVIDE project (working on sustainable urban infrastructures in cities of the Lake Victoria Basin, East Africa), the DESAR project (research and pilot projects in Decentralized Sanitation and Reuse, the Netherlands), and among others within NETSSAF (large scale implementation of sanitation in Africa), and EcoSan networks. The major milestone for this book to emerge was however the IWA Sanitation Challenge Conference of May 2008 in Wageningen, the Netherlands where all the authors of this book presented their papers.

The conference was organized by a consortium of sanitation specialists at Wageningen University's Environmental Policy Group (the editors) and the sub-department of Environmental Technology, LeAF (Lettinga Associates Foundation) and Wetsus (Center of Excellence for Sustainable Water Technology in the Netherlands). It was a unique event as it enabled a truly multi-disciplinary approach in discussing Sanitation Challenges in North and South with social and political scientists, natural scientists, environmental engineers and practitioners in one scientific conference.

This volume presents a selection of the social scientific insights and research results presented at the Sanitation Challenge Conference: the concepts, decision-making support tools and the perspectives from farmers and consumers towards sanitation innovation.

The editors would like to thank all contributing authors for their co-operation and time, and the co-organizers, participants and funders of the Sanitation Challenge Conference. Lastly, we thank Corry Rothuizen at the Environmental Policy Group and Marlies Vlot and Takeesha Moerland-Torpey at Springer Academic Publishers for their help in the final editing and processing of the manuscript.

The editors
Bas van Vliet
Gert Spaargaren
Peter Oosterveer

List of Abbreviations

AT	Appropriate Technology
BOD	Biological Oxygen Demand
BSE	Bovine Spongiform Encephalopathy
C	Carbon
CBO	Community Based Organization
CIDA	Canadian International Development Agency
COD	Chemical Oxygen Demand
CREPA	Centre for Low-Cost Water and Sanitation
CSA	Combined sewer overflow
DANIDA	Danish International Development Agency
DESAR	Decentralized Sanitation and Reuse
DGRE	General Directorate of Water Resources, Burkina Faso
EUWI	European Water Initiative
FC	Faecal Coliform
GDP	Gross Domestic Product
IMF	International Monetary Fund
K	Potassium
KIWASCO	Kisumu Water Supply and Sewerage Company (Kenya)
LCA Life	Cycle Analysis
LDC	Low Developed Country
LG	Local Government
MCDA	Multi Criteria Decision Analysis
MDG	Millennium Development Goal
M&E	Monitoring and Evaluation
MFA	Material Flow Analysis
MIW	Ministry of Water and Irrigation (Kenya)
MM	Modernized Mixtures
MMA	Modernized Mixture Approach
N	Nitrogen
NCCR	Swiss National Centre of Competence in Research
NEMA	National Environment Management Authority
NETSSAF	Network for the Development of Sustainable Approaches for large scale implementation of Sanitation in Africa

NGO	Non Governmental Organization
NWSC	National Water and Sewerage Corporation (Uganda)
ODA	Official Development Assistance
OECD	Organization for Economic Co-operation and Development
O&M	Operation and Maintenance
ONEA	National Office of Water and Sanitation, Burkina Faso
P	Phosphorous
PAQPUD	On-Site Sanitation Program in Dakar's Peri-Urban Areas
PPP	Public-Private Partnership
PrOACT	PRoblem reconnaissance, Objectives, Alternatives Consequences and Trade-off
PROVIDE	Partnership for Research on Viable Environmental Infrastructure Development in East Africa
PSAO	Strategic Sanitation Plan for Ouagadougou
PST	Primary Sedimentation Tanks
SAP	Structural Adjustment Program
SESATS	Semi-centralized Supply and Treatment Systems
STW	Sewage Treatment Works
TF	Trickling Filter
TSS	Total Suspended Solids
VIP	Ventilated Improved Pit (latrine)
UASB	Up-flow Anaerobic Sludge Bed (reactor)
UDDT	Urine Diverting Dry Toilets
UNDP	United Nations Development Program
UNICEF	United Nations Children's Fund
WHO	World Health Organization
WSP	World Bank Water and Sanitation Program

Contents

1 **Introduction** 1
Bas vanVliet, Gert Spaargaren and Peter Oosterveer

Part 1 Social Scientific Concepts of Provisioning Sanitation Services

2 **Meeting Social Challenges in Developing Sustainable Environmental Infrastructures in East African Cities** 11
Peter Oosterveer and Gert Spaargaren

3 **Sense and Sanitation** 31
Bas van Vliet and Gert Spaargaren

4 **Providing Sanitation for the Urban Poor in Uganda** 49
James Okot-Okumu and Peter Oosterveer

Part II Decision-Making Tools

5 **A Flowstream Approach for Sustainable Sanitation Systems** 69
Elizabeth Tilley, Christian Zurbrügg and Christoph Lüthi

6 **A Learning and Decision Methodology for Drainage and Sanitation Improvement in Developing Cities** 87
Joost van Buuren and Astrid Hendriksen

7 **Perceptions of Local Sustainability in Planning Sanitation Projects in West Africa** 105
Jennifer McConville, Jaan-Henrik Kain and Elisabeth Kvarnström

8 **Interactions Between Urban Forms and Source-Separating Sanitation Technologies** 125
Franziska Meinzinger, Volker Ziedorn and Irene Peters

9 **Reconsidering Urban Sewer and Treatment Facilities in East Africa as Interplay of Flows, Networks and Spaces** ... 145
Sammy Letema, Bas van Vliet and Jules B. van Lier

10 **Meeting the Sanitation Challenge in Sub-Saharan Cities: Lessons Learnt from a Financial Perspective** ... 163
Jérémie Toubkiss

Part III Perspectives from Farmers and End-Users

11 **Role of Farmers in Improving the Sustainability of Sanitation Systems** ... 179
Håkan Jönsson, Pernilla Tidåker and Anna Richert Stintzing

12 **Governing Peri-Urban Waste Water Used by Farmers: Implications for Design and Management** ... 189
Reginald Grendelman and Frans Huibers

13 **End User Perspectives on the Transformation of Sanitary Systems** ... 203
Dries Hegger and Bas van Vliet

14 **Conclusion and Discussion** ... 217
Gert Spaargaren, Bas van Vliet and Peter Oosterveer

Index ... 227

Contributors

Joost van Buuren
Sub-department of Environmental Technology, Wageningen University,
Bomenweg 2, 6703 HD Wageningen, The Netherlands
joost.vanbuuren@wur.nl

Reginald Grendelman
Irrigation and Water Engineering Group, Wageningen University,
P.O. Box 47, 6700 AA Wageningen, The Netherlands
reginaldgrendelman@hotmal.com

Dries Hegger
Environmental Policy Group, Wageningen University, Hollandseweg 1, 6706 KN Wageningen, The Netherlands
dries.hegger@wur.nl

Astrid Hendriksen
Environmental Policy Group, Wageningen University,
Hollandseweg 1, 6706 KN Wageningen, The Netherlands
astrid.hendriksen@wur.nl

Frans Huibers
Irrigation and Water Engineering Group, Wageningen University,
P.O. Box 47, 6700 AA Wageningen, The Netherlands
frans.huibers@wur.nl

Hakan Jönsson
Department of Energy and Technology, Swedish University of Agricultural Sciences (SLU), Box 7032, SE-750 07 Uppsala, Sweden
EcoSanRes, Stockholm Environment Institute (SEI), Kräftriket 2b, SE-106 91 Stockholm, Sweden
hakan.jonsson@et.slu.se

Sammy Letema
Department of Environmental Planning and Management,
Kenyatta University, Nairobi, Kenya
sletema@yahoo.com

Jules van Lier
Department of Water Management, Section Sanitary Engineering, University of Technology, Stevinweg 1 Delft, The Netherlands
j.b.vanlier@tudelft.nl

Christoph Lüthi
Eawag: Swiss Federal Institute of Aquatic Science and Technology, Department of Water and Sanitation in Developing Countries (Sandec), Ueberlandstrasse 133, Duebendorf, Switzerland
christoph.luthi@eawag.ch

Jaan-Henrik Kain
Department of Architecture, Chalmers University of Technology, SE-412 96 Göteborg, Sweden
kain@chalmers.se

Elisabeth Kvarnström
Stockholm Environment Institute, Kräftriket 2B, SE-10691 Stockholm, Sweden
elisabeth.kvarnstrom@sei.se

Jennifer McConville
Department of Architecture, Chalmers University of Technology, SE-412 96 Göteborg, Sweden; Stockholm Environment Institute, Kräftriket 2B, SE-10691 Stockholm, Sweden
jenmcc@chalmers.se

Franziska Meinzinger
Institute of Wastewater Management and Water Protection, Hamburg University of Technology, Eissendorfer Str. 42, 21071 Hamburg, Germany
f.meinzinger@tuhh.de

James Okot-Okumu
Makerere University Institute of Environment and Natural Resources (MUIENR), P.O. Box 7298/7062, Kampala, Uganda
jokotokumu@muienr.mak.ac.ug

Peter Oosterveer
Environmental Policy Group, Wageningen University, Hollandseweg 1, 6706 KN Wageningen, The Netherlands
peter.oosterveer@wur.nl

Irene Peters
Urban Technical Infrastructure Systems, Department of Urban Planning, HafenCity University, Schwarzenbergstr. 95, 21073 Hamburg, Germany
irene.peters@hcu-hamburg.de

Gert Spaargaren
Environmental Policy Group, Wageningen University, Hollandseweg 1, 6706 KN Wageningen, The Netherlands
gert.spaargaren@wur.nl

Anna Richert Stintzing
Richert Miljökompetens, Åsögatan 140, SE-116 24 Stockholm
anna@richert.se

Pernilla Tidåker
Svenskt Sigill, SE-105 33 Stockholm
pernilla.tidaker@sigill.lrf.se

Elizabeth Tilley
Eawag: Swiss Federal Institute of Aquatic Science and Technology,
Department of Water and Sanitation in Developing Countries (Sandec),
Ueberlandstrasse 133, Duebendorf, Switzerland
elizabeth.tilley@eawag.ch

Jérémie Toubkiss
Hydroconseil, 198, Chemin d'Avignon, 84470 Châteauneuf de Gadagne, France
toubkiss@hydroconseil.com

Bas van Vliet
Environmental Policy Group, Wageningen University, Hollandseweg 1,
6706 KN Wageningen, The Netherlands
bas.vanvliet@wur.nl

Volker Ziedorn
egeb – Business Development Company Brunsbüttel, Elbehafen,
25541 Brunsbüttel, Germany
ziedorn@egeb.de

Christian Zurbrügg
Eawag: Swiss Federal Institute of Aquatic Science and Technology,
Department of Water and Sanitation in Developing Countries (Sandec),
Ueberlandstrasse 133, Duebendorf, Switzerland
christian.zurbrugg@eawag.ch

Chapter 1
Introduction

Bas vanVliet, Gert Spaargaren and Peter Oosterveer

Abstract This chapter introduces what the social sciences can offer to sanitation and the challenges that lay ahead of providing sustainable access to sanitation to billions of people in both the Western as well as the developing world. It outlines three social scientific perspectives to the sanitation challenge: that of socio-technical change, multi-level governance and consumer studies. The sanitation challenge is three fold: in the Western world the challenge is to initiate a transition from strongly centralized infrastructures towards flexible and sustainable forms, in the developing world to provide access to sanitation for the poor, while the challenge in both worlds is to move beyond traditional dichotomies between small and large, appropriate and modern technologies. Lastly the organization of this volume is explained.

1.1 Sanitation

There is no doubt that meeting the UN Millennium Development Goal of halving, by 2015, the proportion of people without sustainable access to safe drinking water and improved sanitation facilities can be called a challenge. The challenge is even bigger when we also consider those parts of the world that do have access to drinking water and sanitation, but need to modernize the systems they have in order to make them more sustainable.

Before further elaborating on the sanitation challenge, let us first define what sanitation really is about. For environmental health workers and environmental engineers the following definition might have worked for a long time, with the first part satisfying health workers, the latter part the engineers:

> Sanitation is the process of keeping places clean and hygienic, especially by providing a sewage system and a clean water supply. (CollinsCobuild English Dictionary 1995)

B. van Vliet (✉), G. Spaargaren, and P. Oosterveer
Environmental Policy Group, Wageningen University, Hollandseweg 1, 6706 KN Wageningen, The Netherlands
e-mail: bas.vanvliet@wur.nl; gert.spaargaren@wur.nl; peter.oosterveer@wur.nl

B. van Vliet et al. (eds.), *Social Perspectives on the Sanitation Challenge*,
DOI 10.1007/978-90-481-3721-3_1, © Springer Science+Business Media B.V. 2010

But sanitation is of course more than that. Indeed, providing a sewage system and clean water supply has been the dominant technological paradigm for over a century. Between water supply and a sewage system still stands the icon of modernity for the billions of people who are deprived of it: the water closet. Water flush toilets effectively keep places clean and hygienic and facilitate their users the tremendously appealing habit to 'flush-and-forget', leaving water works, municipalities and sewer treatment corporations with the nasty business of transporting and cleaning up human wastes. But slowly policy makers and engineers who have envisaged building water flush toilets for all have come to acknowledge that such endeavor is far from realistic. Sewage systems are extremely costly to build and maintain and spill scarce water for transporting and diluting waste for which smarter uses can be found if it were stored and treated in a concentrated form (Lens et al. 2001). Due to the huge investments needed and consequently its slow uptake and inflexibility, sewage systems can by far not meet the needs of the fast growing urban populations in the developing world. We should start to think of future scenarios for sanitation which do not lead to providing water closets to all as the preferred destination. Many urban and rural settlements around the world probably will never have a centralized sewage system installed. Dry toilets, urine diversion toilets, vacuum toilets, on-site composting, or anaerobic digestion are all potential means to keep places clean and hygienic, while making the need for sewage systems redundant.

Yet it is a common pitfall to stay focused on the diverse toilet and treatment systems of human waste alone. Toilets are a focal point as they are a nod in the sanitation chain and they materialize the relation between users and sanitation service providers. But we should also think about the sanitation chain as a whole. Most commonly a chain is described in either material flows and/or as a series of technological artifacts accommodating such flows. A sanitation flow chain starts with faecal matter and urine, with its organic and mineral compounds, flush water, waste water from various other sources like sinks, washing machines and showers. In many accounts on 'eco-sanitation' or 'resource oriented sanitation' also the reuse of organic matter and minerals in agriculture or the utilization of produced energy in households and industry is included in the flow schemes of sanitation. Instead of a chain, sanitation in these accounts is conceived of as a closed loop. All of these flows are being processed into compost or sludge of organic nature, minerals like phosphate, and gases like methane. Combined or separated, the flows are collected, transported, and treated and reused by a series of technical artifacts or briefly 'technical systems' ranging from large scale sewage and waste water treatment systems to small on-site systems like pit latrines.

We fully acknowledge that sanitation is more than just providing sewage and clean water and definitely more than toilets: eventually sanitation concerns water, energy and nutrient cycles and the technology to facilitate the transport, processing and reusing of flows. But this volume has something else to contribute to the widening discourse on sanitation. We observe that sanitation has so far hardly been addressed from a social scientific perspective. Historians have done their job in describing and explaining the emergence of toilet and sewage systems (Melosi 2000; Van Zon 1986) in Western societies. There are only a few accounts of the social-cultural meaning of toilets (Gastelaars 1996) and in the field of Science and

Technology Studies the study of water supply or water management technologies outnumber by far those of waste water and sanitation technology.

In this volume we aim to fill the gap social sciences have left on conceptualizing sanitation. Like in other systems of provision that have been addressed much more extensively (energy, transport, drinking water supply to mention some), sanitation systems form the material base for a number of crucial social relations and everyday social practices in contemporary societies. The social dimensions of sanitation can be studied firstly on the level of the socio-technical provision of sanitation services; secondly on the level of the governance of sanitation; and lastly on the level of end-users and consumption of sanitation services.

Focusing on the level of sanitation systems of provision, the social sciences have a lot to offer by studying the highly complex interfaces between sanitation technologies, consumers and providers. Left in the hands of technocrats and engineers only, sanitation becomes a narrowly defined problem for which some preferred technological fix can be designed and implemented. The failures of such an approach have come to the foreground everywhere around the globe, but especially in low developed countries. Once sewer systems have been rolled out in ways similar to how the developed world had built their systems, and sunk costs have been made, deviations from water born sanitation routes are hard to achieve. Such path dependencies also obstruct innovative routes to be taken in developing countries. In many African cities the heritage of a colonial sewer system is still limiting the innovation paths for wiser, more flexible sanitation options for the urban poor, even in cities where sewer connections amount for only a few percent of the city population. On the other hand, proponents of alternative sanitation solutions like 'eco-sanitation' or 'decentralized sanitation and reuse' show a tendency to develop belief systems that do not allow for large scale technological routes, or possible combinations with the so much disputed large technical sewage systems. Leaving aside the grown dichotomy between small, decentralized sanitation solutions on the one hand and large scale sewage systems on the other, this book aims to provide insights in what we have labeled 'Modernized mixtures' of social and technical scales in sanitation. The theoretical and empirical elaboration of the concept of Modernized Mixtures is one of the leading themes in this volume.

Second, we observe various dynamics in the governance of sanitation service provision nowadays. Municipal or state organizations that for decades have had a monopoly in providing water and sewer services start to dissolve and give way to various forms of private participation in sanitation service provision. This is a world-wide tendency. Private participation means the involvement of the private sector at all scales: from multi-national utility companies to small local firms as well as local associations of ultimate users of sanitation services. Such differentiation in provisioning actors gives way to a range of new sanitation services to be tested and implemented. Decision-making on sanitation is increasingly opened up, no longer are the development paths being outlined at forehand. It is time to develop tools to enable the decision-making on sanitation by a much wider range of actors than state-actors only, and at multiple levels of society. This is the second ambition of this volume concerning the social perspectives on the sanitations challenge: to provide insights in new decision-making tools for sanitation in the developed and developing world.

Lastly, sanitation is a crucial but mostly hidden aspect of everyday life and consumption. While the black boxes of other infrastructure-related consumption (water, domestic waste, energy, transport) have been opened up recently (Southerton et al. 2004; Shove 1997; Spaargaren and Van Vliet 2000; Van Vliet 2002) analyzing in some detail how captive consumers can be encouraged to make lifestyle choices about the way they consume or even produce their own water, electricity, waste or transport services, this has not been the case for sanitation, until very recently (Hegger 2007). In terms of enabling conspicuous consumption of infra-related services, quite a lot has changed over the past 10–20 years. Excessive energy or water use or – on the contrary – excessive savings realized on both are being displayed to the wider society with fashionable living rooms, bathrooms, kitchens and their high-tech appliances. Abundant are the magazines and TV shows on how and where to refurnish, to bath, to cook and to dine. However, the toilet seems to remain a last resort of private life, out of sight and out of fashion. And what counts for highly developed consumerist societies seems to be as valid for low developed countries in this respect. Toilets and sanitary practices are sealed-off from society and culturally loaded with negative notions of dirt, shame and waste. Yet toilets are the key to realizing the Millennium Development Goals and improving the living conditions of billions of people. In the developed world flush toilets represent the biggest share of domestic water usage and waste water production. If this environmental burden is ever to be changed into more sustainable directions, we need to understand how standards of cleanliness, convenience and hygiene have been constructed over time and what strategies are needed to initiate a trend in a default direction. A third aim of this volume is thus to provide an end-user perspective on innovations in sanitation.

1.2 Three Worldwide Sanitation Challenges

We believe that social perspectives on sanitation as outlined above are needed to address the various challenges that the world is facing in terms of sanitation. The sanitation challenge in the Western world is to initiate a transition from strongly centralized, water-based infrastructure regimes towards more sustainable sanitation regimes. Perhaps even more than in other infrastructures there is a huge inertia throughout the chain of sanitation. Inertia because of the sunk costs made in sewage infrastructures, in the interests of providers, in the minds of engineers and last but not least, in consumer and end-user habits when it comes to sanitation. A transition in sanitation therefore entails major efforts at multiple levels and by a wide range of actors. But where to start? A great deal of current research on innovation in sanitation is built around pilot projects in which new technologies are being tested. In many cases they do not more than just that: testing the technological feasibility of a new toilet or treatment unit. But increasingly other fields of testing are being introduced at pilot projects: evaluation of consumer responses, of various decision-making procedures, of the willingness of chain actors to reuse waste waters, urine or other products derived from new sanitation technologies. Not surprisingly, niche management and transition management approaches have become favorable among social scientists to study the possibilities of traditional technological pilots in sanitation. Such approaches offer the tools to study processes of social and institutional learning, or the formation of new sanitation regimes. Throughout this book the experiences of a range of sanitation pilot projects form the empirical basis for further social scientific exploration of consumer perspectives, system transformation and governance issues.

In the developing world the sanitation challenge is to provide improved sanitation services to the poor and the very poor, without compromising on sustainability. New configurations of employing the best practices of sanitation technology and management for rural and urban contexts need to be explored and evaluated. Also here a break-through is needed in decision-making about sanitation, be it of a different kind and at a different level when compared to the developed world. In the absence or at least severe shortage of water and sanitation infrastructures, like in the wide spread urban slums of third world cities, the room for any decision-making about the various sanitation options appears to be incredibly wide. But in fact it is restricted by lack of funding, political commitment and above all, a lack of imagination. All too often standard models of modern sanitation are being portrayed, with a sewage system and water flush toilets as an ultimate stage in a process of modernization. Or, alternatively, eco-sanitation options are envisaged that are, in similar vein, unaffordable or unacceptable for the mass of users addressed. The contributions in this volume addressing the sanitation challenges of the South all try to stimulate our imagination about other routes to take, either by providing new concepts, new decision-making tools or new insights in the perspectives from users of sanitation systems.

Eventually the sanitation challenge in both the Western and developing world is to go beyond traditional dichotomies between 'small, appropriate' and 'modern, advanced' technologies and to develop rural and urban sanitation with a mix of scales, strategies, technologies, payment systems and decision-making structures, that better fit the physical and human systems for which they are designed. New concepts are needed to bridge these gaps. Proposals presented in this volume include that of 'Modernized Mixtures' introducing various social and technical variables in assessing and deciding on sanitation options and a 'flowstream approach', with proposals to organize and define the diverse elements in sanitation flow systems.

1.3 Organization of the Book

This book presents a selection of revised and extended papers presented at the international IWA Conference called "Sanitation Challenge", in May 2008 at Wageningen University. To obtain a full picture of contemporary social perspectives on the sanitation challenge we have grouped the chapters in three sections, addressing new concepts, new decision-making tools and perspectives of farmers and end-users respectively. With these three themes we cross-cut the different levels of social scientific research outlined above: that of Science and Technology Studies, Multi-Level Governance, and of Consumer Studies. Most chapters however address a combination of such approaches and combine theoretical outlines with empirical findings.

1.3.1 Part 1 Social Scientific Concepts of Provisioning Sanitation Services

The contribution of Oosterveer and Spaargaren (Chapter 2) presents a new socio-technical framework, called 'Modernized Mixtures', in which solutions for sanitation problems are sought in diverse mixtures of social and technical measures at different scales. The framework has been applied in recent sanitation research in Eastern Africa and in Western Europe.

In Chapter 3, 'Sense and Sanitation' Van Vliet and Spaargaren analyze the implications for sanitation infrastructures of a tendency to re-visualize and re-sensitize infrastructural services. Case studies are used to present a toolbox for analyzing the opportunities of the ambivalent relationship between sanitation and the senses.

Finally in Chapter 4, Okot-Okumu and Oosterveer provide an analysis of the problems of sanitation services provision in urban centres of Uganda. It leads to a set of decision-making priorities: especially to provide improved sanitation services to the peri-urban poor.

1.3.2 Part II Decision-Making Tools

The section on decision-making tools is opened with Chapter 5 by Tiley, Zurbrügg and Lüthi presenting a flowstream approach to sanitation: a novel method for organizing and defining sanitation systems. Its advantages and shortcomings are being discussed and a proposal to standardize the way in which sanitation is thought of communicated.

Van Buuren and Hendriksen present in Chapter 6 a method to support decisions about drainage and sanitation systems based on multi-criteria analysis in combination with stakeholder dialogues. It is supported by a data base of 52 drainage and sanitation options, assessment criteria and a performance matrix. Pilot workshops organized in Ho Chi Minh City (Vietnam) and Kampala (Uganda) offer insights in the application possibilities of the toolbox.

Based on interviews with local actors, McConville, Kain and Kvarnström present in Chapter 7 criteria for sanitation interventions in West Africa and compare them with those derived from international literature. Criteria from the field relate much more to institutional and behavioral aspects, while international literature refers more to technical, health and sustainability issues. The solution to the sanitation challenge requires understanding and merging of these perspectives into a comprehensive approach.

Meinzinger, Ziedorn and Peters examine in Chapter 8 how specific characteristics of the urban form can impact on the implementation of new sanitation technologies. A typology for residential areas in Germany based on physical characteristics supports the definition of four different source-separating sanitation systems. Assessments are made about feasibility and cost-effectiveness of the systems in different areas of Hamburg.

Diverse patterns of modernization have resulted in a variety of sanitation systems in urban East Africa. Letema, Van Vliet and Van Lier assess in Chapter 9 the different systems in terms of flows, networks and spaces, which serves as an input to analyze the opportunities for so-called modernized mixture systems of sanitation in Kisumu (Kenya) and Kampala (Uganda).

Finally in Chapter 10 Toubkiss addresses the need for novel finance structures for sanitation in sub-Saharan Africa and evaluates the findings of diverse projects in which these systems have been applied.

1.3.3 Part III Perspectives from Farmers and End-Users

Jönsson, Tidåker and Stintzing argue in Chapter 11 that farmers are just as essential as toilet users to recycle the nutrients of excreta in a safe and resource efficient way. Farmers should therefore participate in the initial planning of recycling sanitation systems. Swedish experiences are presented on how to include farmers in the organization of applying the recycled product on the farmland.

Grendelman and Huibers approach sanitation as a reversed waste water chain in Chapter 12. It challenges environmental engineers to incorporate downstream farmers demands and necessities in the design of waste water facilities as to provide them the right mix of nutrients for waste water irrigation. Peri-urban farmers should therefore be involved in what can be called agricultural waste water management.

Lastly, looking from the other end of the sanitation chain, Hegger and Van Vliet in Chapter 13 present a toolkit to better understand and experiment with end-user roles in sanitation innovation, based on a study of pilot projects in Western Europe. They propose and outline an end-user perspective that is needed to get demonstration projects realized and to contribute to successful innovation routes in sanitation.

The conclusion and discussion presented by Spaargaren, van Vliet and Oosterveer in Chapter 14 draws upon the common themes reappearing in all chapters: the need for integrated, socio-technical solutions, based on local situations and knowledge. The chapter makes up the balance of the book and provides a future agenda of research and policy making on sanitation systems and their end-users.

References

Gastelaars, M. (1996). The water closet: Public and private meanings. *Science as Culture, 5*(4), 483–505.

Hegger, D. (2007). *Greening sanitary systems: an end-user perspective*. Ph.D. thesis, Wageningen University, Wageningen.

Lens, P., Zeeman, G., & Lettinga, G. (eds). (2001). *Decentralized sanitation and reuse: concepts, systems and implementation*. London: IWA Publishing.
Melosi, M. (2000). *The sanitary city. Urban infrastructure in America from colonial times to the present*. Baltimore, MD: The Johns Hopkins University Press.
Shove, E. (1997). Revealing the invisible: sociology, energy and the environment. In M. Redclift & G. Woodgate (Eds.), *The international handbook of environmental sociology* (pp. 261–273). Cheltenham: Edward Elgar.
Southerton, D., Chappells, H., & Van Vliet, B. (eds). (2004). *Sustainable consumption: the implications of changing infrastructures of provision*. Cheltenham: Edward Elgar.
Spaargaren, G., & Van Vliet, B. (2000). Lifestyles, consumption and the environment: The ecological modernisation of domestic consumption. *Environmental Politics, 9*(1), 50–76.
Van Vliet, B. (2002). *Greening the grid: the ecological modernisation of network-bound systems*. Ph.D. thesis, Wageningen University, Wageningen.
Van Zon, H. (1986). *Een zeer onfrisse geschiedenis: studies over niet – industriële vervuiling in Nederland, 1850–1920* (A very dirty affair: studies in non – industrial pollution in The Netherlands, 1850–1920). Ph.D. thesis, Groningen University, Groningen.

Part I
Social Scientific Concepts of Provisioning Sanitation Services

Chapter 2
Meeting Social Challenges in Developing Sustainable Environmental Infrastructures in East African Cities

Peter Oosterveer and Gert Spaargaren

Abstract The slum population in sub-Saharan Africa is expected to grow from 101 million in 1990 to 313 million in 2015. Modernizing sanitation therefore has to adapt to the context of cities with high densities of poor people under the conditions of absent or fragmented environmental infrastructures and services. Addressing this problem requires an integrated approach that deviates both from the Western large-scale, high-technological, and grid-based systems, as well as from the small-scale, low-tech, decentralized alternative options. A Modernized Mixtures approach should be developed that combines the strong elements from these opposing alternatives. This chapter presents the Modernized Mixtures approach and its contribution to sustainability. It discusses the contribution this approach can make to improving accessibility of urban infrastructures for the poor, while strengthening flexibility and resilience. It is argued that the successful introduction of a Modernized Mixtures approach to urban environmental infrastructures in East African cities requires the careful consideration of social and political factors next to technological innovation.

2.1 Introduction

Africa is going through a process of rapid urbanization. Whereas in the early 1900s, 95% of Africa's population was rural, by 2010 at least 43% of the population will be urbanized (Boadi et al. 2005). The large majority of these new urbanites lives in unplanned, or informal, settlements and therefore the slum population of sub-Saharan Africa is expected to grow from 101 million in 1990 to 313 million in 2015 (Kombe 2005; UNDP 2005). These rapid changes in housing practices signify serious challenges for these people themselves as well as for municipal authorities.

Municipal authorities are faced with the task to provide the expanding populations with adequate infrastructures and services for water, sanitation and solid waste.

P. Oosterveer (✉) and G. Spaargaren
Environmental Policy Group, Wageningen University, Hollandseweg 1, 6706 KN, Wageningen, The Netherlands
e-mail: peter.oosterveer@wur.nl; gert.spaargaren@wur.nl

B. van Vliet et al. (eds.), *Social Perspectives on the Sanitation Challenge*,
DOI 10.1007/978-90-481-3721-3_2, © Springer Science+Business Media B.V. 2010

Assuring these facilities is particularly difficult because they target predominantly low-income settlements with high population densities and high illiteracy rates, under very low living and livelihood security standards, where formal property rights and land titles are absent and existing infrastructures and access to other social services are rather poor. Currently most people in the larger African cities are forced to assure their access to environmental facilities themselves, often against high costs. The resulting poor and inadequate water provision, failing solid waste management and incomplete sanitation facilities result in health hazards and waste materials polluting water, soil, and air. Well-managed systems for piped water, sanitation, drainage, and garbage removal would greatly diminish the health hazards to which city residents are currently exposed and reduce their poverty, even without increasing their income (Satterthwaite 2004). Creating reliable urban environmental infrastructures and services is therefore recognized as of key importance by the African governments and within international development cooperation as expressed in the seventh Millennium Development Goal (MDG) (UN-Habitat 2006). Realizing this shared objective in a sustainable manner is however complicated. The absence of formal governmental interference and the unplanned nature of many neighborhoods, necessitates radical adaptations in technologies, socio-economic management arrangements, and governmental policies. This chapter elaborates on the challenges and opportunities to provide sustainable sanitation and solid waste services in informal settlements under these conditions and thereby makes use of recent research in Africa, particular in urban centres of East Africa (Kenya, Tanzania and Uganda). This is done from a social science perspective. First, we discuss in some more detail the problem of urban environmental infrastructure provision under situations of low living and livelihood standards. We argue for the need to develop an approach to infrastructure provision which better fits the local conditions in both social and technological respects. This so-called Modernized Mixtures approach is illustrated with some examples taken from an ongoing research project on environmental infrastructures in East Africa.[1] Particular attention is paid to the political (planning) aspects of Modernized Mixtures at different levels of scale. We conclude by discussing the potentials of the suggested Modernized Mixtures approach for research and policy making on environmental infrastructures.

2.2 Two Opposing Views on Urban Environmental Infrastructures

Over time, particularly since the nineteenth century Industrial Revolution, many European and other Western countries have successfully established reliable urban environmental infrastructures through installing large centralized systems (Guy

[1] This research is undertaken within the framework of the PROVIDE project. Funded by Wageningen University, PROVIDE focuses on and contributes to the improvement of sanitation and solid waste management in East Africa (Kenya, Uganda and Tanzania) with an emphasis on the Lake Victoria Region. See: www.provideafrica.org

and Marvin 1996). Most towns and villages in OECD countries currently have a well-developed sewage system as a central grid connecting most houses to a waste water treatment plant for processing and discharging waste. All connected households have to pay substantial fees to contribute to the development and maintenance of these systems, while keeping up the quality standards and provisions which are laid down in governmental laws and urban by laws. Solid waste collection and treatment systems follow a similar logic of centralized organization and treatment, mostly ending in sanitary landfills or large incineration plants.

Despite their intention to copy this example, most cities in developing countries have not achieved such a modernization of their facilities. Urbanization under conditions of poverty has given rise to peculiar urban land development patterns, that is informal settlements, that are defying spatial planning theories used to applying master and structure plans (Kombe 2005). Such informal settlements are the result of a dynamic, largely self-managed land development process based on social trust and support mechanisms from social actors at the grass roots. Urban authorities have been far less successful in implementing large infrastructural systems than in OECD countries as they face a number of pertinent and persistent problems, in particular the lack of adequate material and human resources but also specific ecological, institutional, political and cultural challenges (see Box 2.1). Though some cities possess a central sewerage system

Box 2.1 The history of sanitary infrastructure in Uganda

The British completed the first central sewerage system in Kampala by 1939, which included 35 miles of sewers and 27 miles of storm water drains (Nilsson 2006). Development of piped water supply also started during the colonial period in the 1940s. The majority of the systems were constructed from 1950 to 1965, mainly to serve the workers and the small commercial communities. In Uganda in the 1960s sanitation and environmental health were well supported and latrine coverage was high (90–96%). At the time the urban population was much smaller than today. In the 1970s and early 1980s the political turmoil and breakdown of law and order reduced latrine coverage to 30% in 1983. No new schemes were constructed between 1965 and 1990. Only maintenance of the existing schemes was done, but even this was poor. By 1990 virtually the whole urban water infrastructure was run down and serving less than 10% of the population in the large towns.

In the late 1980s a fresh effort was made to accelerate the promotion of sanitation from new projects. The National Water and Sewerage Corporation was established as a government parastatal with a mandate to operate and provide water and sewerage services in areas entrusted to it on a sound commercial and viable basis. The rest of the water supplies were operated by the Directorate of Water Development. By 1990, there were only 37 urban water systems including those under the NWSC (Mubeezi 2007).

dating from the colonial era, which mostly only covers the central business district and some richer areas, even maintaining these often overburdened systems constitutes a challenge. Paying for the necessary investment and operational costs of such infrastructures bleeds money out of the social system and runs counter to catering for other local needs.

In response to these limitations in government-managed systems, alternative approaches are suggested and particularly privatization is repeatedly considered a solution, as private companies are expected to be more efficient and more responsive to client demands. However, the pressure to introduce (full) cost recovery for collection and processing and for managing sanitary infrastructures forces private companies to seek rents from serving (only) the higher income areas or fully paid services, leaving poor and marginal areas under the responsibility of under-resourced local authorities and Community Based Organizations (CBOs). However, whether publicly or privately managed, many of the large scale infrastructural systems in developing countries prove to be limitedly resistant against the political, economic, ecological and social instabilities they face, leading to poor environmental performances and perpetual breakdowns, due to lack of proper maintenance or timely investments.

In response to these problems with large centralized environmental infrastructures, alternative solutions, particularly small-scale and decentralized systems (Schumacher 1973), are proposed. Decentralized sanitation and reuse (DeSaR) systems were developed partly in opposition to centralized ones (Mels et al. 2005) and claim to be more robust, cheaper and better able to deal effectively with environmental challenges like high levels of water consumption and indiscriminate discharge of potentially valuable substances (Lens et al. 2001, 2003; Otterpohl et al. 1997).[2] Improved pit latrines, small-scale household composting and other decentralized systems for reuse of solid waste are widely considered a potential solution for developing countries. While this appropriate technology (AT-)paradigm has certainly booked major results, it remained being considered a simple, second-best technology paradigm, useful in situations where the finances, technological capabilities and organizational capacities were severely limited. Where introduced, these technologies offer solutions for individual households but they do not solve the massive challenges of addressing the sanitation challenges of large cities in developing countries. Both users and local authorities consider such technologies 'low quality'. In practice, these first generation decentralized, 'appropriate' technologies are being replaced with more advanced systems as soon as the social, economic and technological conditions allow for it. Large cities in developing regions such as East Africa are therefore faced with the dilemma of which path to choose for improving their sanitation and solid waste infrastructures as both large-scale centralized systems and small-scale decentralized systems each show serious weaknesses.

---

[2] See also Hukka and Katko (2003), Mistra (2002), Seppälä et al. (2004), Van Vliet (2006) and Hegger (2007) for attempts to apply DeSaR technologies in urban settings.

2.3 Modernized Mixtures for Improving Urban Environmental Infrastructures

Providing sanitation and solid waste services, if this is to contribute to improving daily lives of the urban poor in East Africa, has to be adapted to the context of cities with high population densities, with people living in informal communities where environmental infrastructures are fragmented or completely absent.

For this, the first generation of decentralized systems (appropriate technologies) has to be replaced with a second generation that offers more sustainable alternatives both from technical and from social points of view. We suggest the concept of Modernized Mixtures (Spaargaren et al. 2006) referring to the development of systems which 'build upon' decentralized units of the DeSaR-type but which try to create solutions at a larger scale and take into account the specific local conditions of developing countries. Applying such a Modernized Mixtures (MM) approach to sustainable urban development means the introduction of an 'organized eclecticism' by combining various levels of scale, strategies, technologies, payment systems and decision-making structures, to create a better fit with the physical and human systems for which they are designed. This approach is referred to as 'mixture' because it takes the best features out of both (modern) decentralized and centralized systems, and combines them into hybrid solutions which better fit the local situation.

When working with MM, one leaves behind the (essentially) false dichotomy dividing centralized, large-scale, high-tech solutions from decentralized, appropriate, small scale and low-cost technology solutions. Instead, the best of both paradigms has to be combined into configurations that represent the low cost, accessible and robust performance of decentralized systems while at the same time realizing the economies of scales and high urban density-capacity characterizing centralized systems. DeSaR-like systems have turned out to be performing best in close relationship with or even in certain dependency from (elements of) conventional large scale socio-technical systems (Lens et al. 2001, 2003; Van Vliet 2006).

Figures 2.1 and 2.2 together illustrate the basic notions of the MM approach, bringing together elements from both paradigms in a number of options and strategies, adapted to the particular infrastructural, institutional, economic and ecological contexts. Figure 2.1 represents the relevant dimensions that have to be taken into account when developing urban environmental infrastructures for water and waste(water) services, while Fig. 2.2 illustrates some possible ways in which these dimensions can be combined into specific Modernized Mixtures.

By moving towards the upper-right corner in the model, infrastructures tend to resemble the large scale publicly-managed, central grid-based systems in industrialized, developed countries. Moving to the bottom-left corner in this model visualizes the decentralized, small-scale systems (like EcoSan) developed in the past for developing countries as well as particular DeSaR-solutions for industrialized countries. The third example shows different Modernized Mixtures, adapted to the specific local contexts and requirements. Integrating knowledge with respect to all relevant dimensions is needed to optimize the chances for socio-technical systems to fit into the specific local social and technical conditions.

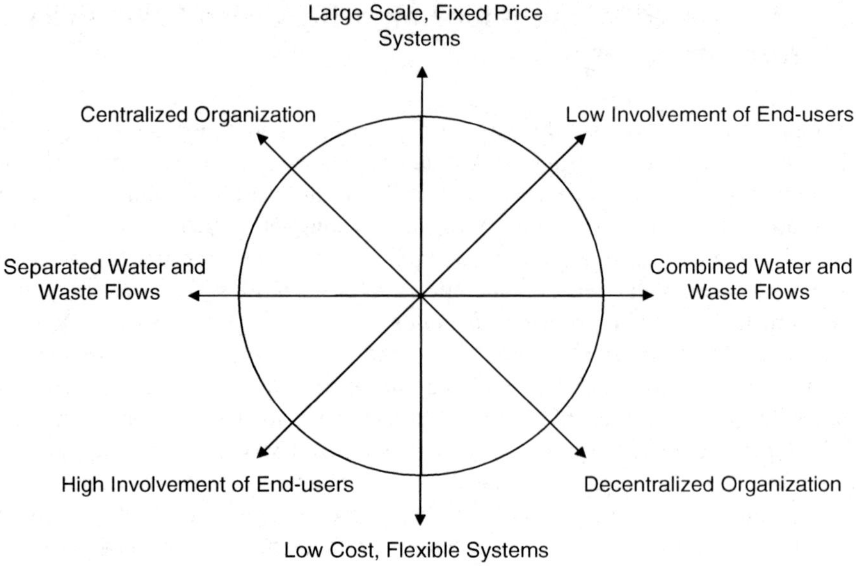

Fig. 2.1 Dimensions of environmental infrastructures (Spaargaren et al. 2006)

This new, hybrid paradigm can be characterized by its multidimensional character in technological (scale, process and combination of water and waste flows) and management/governance respects (involvement of end-users, financial arrangements and organizational set-up). In addition, MM opts for an integrated approach, including all steps in the urban solid waste and the sanitation chains, combining multiple scales in organization, management and governance, and requiring the inclusion of technical as well as social scientific knowledge. The objective of introducing MM is to create a 'fit' between different infrastructural options and the prevailing socio-economic, ecological, technological, and political conditions. For realizing this objective one has to develop and built upon a profound understanding of the specific (urban or semi-urban) settings in which these infrastructures are to be realized. This understanding will allow the involved stakeholders to answer the question, which technological options (and combinations) are realistically possible in the context of their particular cities. This means that each city, or even each neighborhood, will require a specific mixture of technologies and institutional arrangements. Hence, promoting MM means promoting a modular approach to urban environmental facilities and not a one-size-fits-all solution.

In order to provide adequate solutions, MM approaches should be ecologically and institutionally sustainable, accessible (particularly for the poor), and institutionally and technically flexible, resilient and robust. These criteria are all relevant but their exact meaning in the context of particular cities and their relative weight cannot be determined beforehand. Nevertheless some further thought on their connotation can be helpful here. Accessibility reflects the extent to which specific groups within the urban population, such as women, poor or elderly, are included

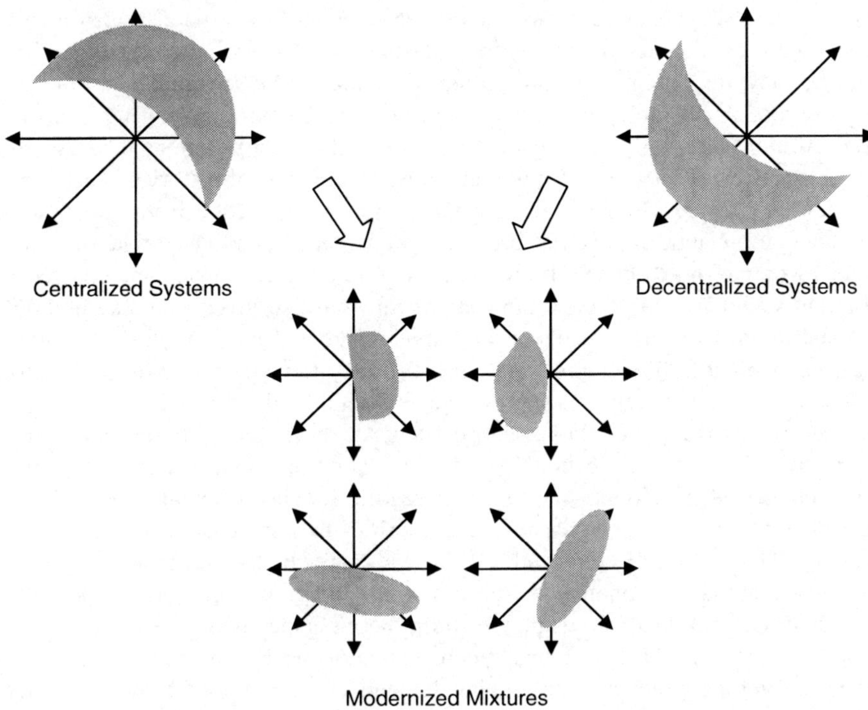

Fig. 2.2 Modernized mixtures (MM) as alternative to centralized and de-centralized systems (Spaargaren et al. 2006)

or excluded from receiving sanitary infrastructures and services due to financial, physical, or cultural reasons. The flexibility criterion points at the way a sanitation system might fit into more encompassing systems to be developed in the future, while also describing how the systems behave in times of instability of climatic, political, or economic nature. The sustainability requirement can be distinguished in institutional and ecological sustainability. Institutional sustainability concerns the extent to which a new system becomes embedded in existing socio-political and cultural systems at the local and national level, while improving their performance. Ecological sustainability refers to the achievements in waste prevention (reducing the need for final disposal of the waste) and reducing the demand for inputs, in particular water and energy. The criteria should be further developed through stakeholder workshops to elaborate the particular connotation in the specific context and their relative importance enabling a comparison between different options (See Van Buuren and Hendriksen in Chapter 6 of this volume). This way a matrix can be developed, where criteria are set and indicators for measurement applied.

Introducing MM may be complicated as existing systems and practices are stabilized through multiple levels of scale and numerous social actors, thereby creating various ‘lock-in’ effects (cf. Geels 2004, 2005, 2006; Geels et al. 2004).

Socio-technical systems such as urban environmental infrastructures also involve larger regimes, that is shared routines or rules that co-ordinate activities. Such regimes may, for example, include technical standards and governmental rules that favor particular technologies, policies, or cultures. Resource management institutions are not just formalized, visible entities but also manifestations of negotiated social practices, located both locally and in wider contexts of history and economy. In addition, people differ in their capacity to shape collective institutions; some command more authoritative resources and are better placed to negotiate rules than others. Regimes may change through different trajectories, carried and enacted by different social groups. Radical innovations may have survived as niches and only become mainstream as a result of mal-adjustments and tensions in the dominant regimes (Walker 2000). Innovations may also be enforced by one or more actors within the dominant regime, such as governments and private companies. The absence of 'conventional' large-scale systems in most East African cities gives alternative, decentralized technologies for sanitation and solid waste management a fair chance of developing into a wider regime for the sustainable provision of environmental services, as pressure on authorities to solve these urgent matters is growing (Geels 2005; Kemp et al. 1998). Alternatively, the limitations in locally available material and financial resources in combination with a high dependency from international donor funds may restrain these perspectives.

When suggesting MM-solutions, specific attention needs to be given to the importance of combining technical and social elements in identifying the correct mixture of options. Urban environmental infrastructures and services are particular because they involve different dimensions and different levels of scale. Environmental infrastructures obviously have a technological dimension, but they need to be implemented and managed in order to fulfill their task. At the same time they also need to accommodate the (sometimes widely diverging) local cultural practices and perform in a sustainable manner to prevent negative environmental impacts. Furthermore, managing these infrastructures effectively also requires some form of coordination between institutions and actors at multiple scales: neighborhood, city, national-level, and sometimes even the global level. The environmental flows approach (Mol and Spaargaren 2006) suggests a way to integrate these different dimensions when addressing urban environmental infrastructures. This approach studies the material flows, such as waste or water, and their interaction with the relevant socio-cultural, institutional and ecological environment starting from their generation, via their collection, treatment, eventual reuse, up to their final discharge. These material flows are channeled through networks and scapes together with non-material flows of money, information and people. Both material and non-material flows are constitutive for the environmental infrastructures.

When applying such an (environmental) flows perspective to urban infrastructure provision it is important to note that the networks and scapes organizing the different flows are not homogeneous in character. In the context of solid waste and (waste) water chains the distinction between up-stream and down-stream actors, processes and relationships is of particular importance. Material flows should be approached in an integrated way, taking into account the process from generation to final disposal of (solid) wastes. While traveling through (sanitary) chains and networks however, these flows connect fundamentally different social practices (Giddens 1984). Social actors and institutions involved in the generation and primary collection of sanitation and solid waste (down-stream) differ in some essential ways from those involved in practices of secondary collection, treatment and disposal (up-stream). While the first, down-stream phase is dominated by domestic rationalities and household practices, the second, up-stream phase is primarily characterized by system rationalities of technological and economic nature. Acknowledging the existence of these different rationalities makes the point or location where both rationalities meet of particularly importance from a theoretical and policy perspective.[3] At this location, the different practices and their associated rationalities have to become mutually adjusted and accommodated in order to make possible a more effective, integrated operation and management of infrastructural systems and their environmental flows.

In East Africa, the MM approach may originate from already existing local solutions which then will have to be considered as potential building blocks for larger systems and dealt with within the frame of the complete waste-water-chain, from the local up to the regional and international levels. These considerations may serve as guidance for assessing the feasibility of Modernized Mixtures.

2.4 Perspectives for a Modernized Mixture-Approach in Urban Centres in East Africa

The absence of large sewerage systems, the unsustainability of most existing decentralized installations in combination with the rapid population growth make the task of improving urban sanitary infrastructures in East Africa a pressing need. Under these conditions, the Modernized Mixtures perspective (MM) may serve as a useful guide. In this section we further concretize the MM perspective by discussing some findings both from the literature on environmental infrastructures and from different case-studies in East Africa.[4]

When compared to the situation in many OECD countries, the existing environmental infrastructures in East African urban centres do not contain many lock-in effects for alternative approaches. The most widely used sanitary facilities in poor neighborhoods are pit latrines and they are occasionally supplemented with flushing toilets and septic tanks (Sano 2007). Such conventional pit latrines constitute a traditional and cheap way to handle human waste and require little maintenance. However, they do not provide much comfort, attract flies and can be a source of

---

[3] Examples of such a location – referred to in the literature as consumption junction (Schwartz-Cowan 1987) – are toilets in general or skips, or containers, used in many African cities as collection points for domestic solid wastes.

[4] The case-studies were performed in the context of the PROVIDE project. See footnote 1.

various diseases, so users will not necessarily hang on to them. On the other hand, the automatic flushing toilets connected to septic tanks which constitute the preferred system for most citizens as this option offers comfort comparable with Western cities, are faced with problems as well. Septic tanks are expensive and therefore not affordable for the majority of the urban population and they cease to work properly when they are old and over-utilized, potentially causing serious environmental and public health problems. As for the management of urban solid wastes, the situation is not very different: separate waste collection schemes are almost absent, waste littering is omnipresent, and infrastructures for channeling the waste flows are fragmented and underdeveloped. Since the currently applied technologies face serious challenges, there is room for introducing novel options.

Appreciating these novel options requires an integrated flows perspective from waste generation to discharge, combining social and technological dimensions and accommodating different levels of scale. Our intention here is not to present one fixed and fully elaborated alternative model, but to present experiences and views that may provide a basis for introducing varied sanitary systems and services that are flexible, accessible, and sustainable. As for example Kyessi (2005) has illustrated for the case of water provision, it is important that technological options should not be developed as part of fixed models, but with a special eye on the poor, as an important element in creating different, flexible models with enough room for progressive improvements. In what follows we present experiences and views concerning the development of sanitary systems at four levels: household, community, city and national level. For all these we focus in particular on the socio-cultural and political dimensions of the MM's to be developed in East Africa.

2.4.1 Households and Their Socio-Cultural Norms

The main end-users of sanitation facilities and services are located in households where people live and sometimes also work together. Understanding household dynamics and their interaction with the other elements in the waste and sanitation systems is therefore essential. Targeting the urban poor requires understanding their way of life, including their particular culture, household-composition and dynamics, food-security and income-generating strategies in combination with the relevant formal and informal institutional settings. Among these household dynamics we find tenure arrangements and cultural dynamics as particularly relevant, as will be further elaborated below.

Informal or irregular settlements comprise between 30% and 70% of the population in the large cities of the developing world and up to 85% of the new housing stock is produced in an extra-legal manner. The present statutory and customary tenure systems fail to meet the needs of the lower-income groups so traditional practices of land delivery persist and an organic process of human settlement evolution continues. In most African countries, the currently existing institutions of land management were inherited from the colonial era and have, for various reasons, undergone little modification to reflect changing circumstances. As a result, they have been described

Box 2.2 Access to land for housing in Kampala, Uganda

In Kampala, access to land has predominantly relied on the initiative of the households concerned, often undertaken outside the minimally enforced state regulator framework. These households have to some extent used the niches created by the complexity of tenure rules in the city and by administrative turmoil, in order to informally access land for shelter. Most households do not have official titles because processing a title is a lengthy, cumbersome and expensive venture. Moreover, most families in informal neighborhoods settle on marginal land, which is legally inalienable for development and cannot be titled at all. Furthermore, most landholdings in these informal settlements are too small according to the legal standards. Kampala City Council requires that plots be surveyed only in blocks, to allow some planning of the neighborhood. However, given the nature of sub-divisions in these informal settlements this is hardly possible. In addition, for those who buy land from occupants (as opposed to registered owners), a second payment would have to be made to the titleholder before the latter could sign the transfer form necessary for processing the title. This is, in essence, paying for the land twice and thus a disincentive to land registration (Nkurunziza 2007).

by many as inappropriate, alien, expensive and cumbersome. Formal regulations have not been effective in replacing customary ways of accessing land and housing. So, in reality formal and informal institutions exist next to each other; poor households borrow from or utilize state rules where appropriate and circumvent them when they consider them unaffordable or retrogressive. See Box 2.2 for an exemplary case on Kampala, illustrating the complexities surrounding access to land.

When it comes to developing sanitation infrastructures, particular cultural dynamics tend to be ignored, although they turn out to be highly relevant for the acceptance and ways of use by households. For instance, Scheelbeek (2006) found that the numerous Muslim families in Mwanza (Tanzania) make particular demands on the design of sanitary facilities. According to their norms and believes these facilities should meet several prerequisites:

- The toilet should not face the eastern direction (or kibla).
- There should always be water available for cleaning body parts in close vicinity of the toilet itself.
- Flush toilets should be provided with a lid to avoid water splashing out of the toilet; if a person gets in contact with splashed water, the whole body has to be washed to consider it again as holy.
- The toilet should be separated from the bathroom; this implies a double drainage pipe for both toilet and bathroom.

Designers and developers of sanitation systems cannot ignore these and other cultural prescriptions which orient household members in the way they assess and make use of these systems. This example and many other experiences also make clear that cultural beliefs and practices are very much ingrained in the everyday lives of people and cannot simply be adapted to technological requirements through information campaigns or awareness raising. For this reason, designers and managers should involve the community in their decision-making process, while acknowledging the existing differences within communities between men and women, between different cultural and ethnic groups and between children, adults and the elderly.

Existing social practices, norms and values of end-users should be taken into account, as particular interventions may be unacceptable and need to be adapted, or replaced and complemented by other measures or techniques. For example, urine separation by the implementation of sitting-down toilets will not be supported within Muslim communities. In a case study on solid waste, the importance and variability of on-site storage was underlined. Households in Dar es Salaam (Tanzania) have no access to standard domestic storage facilities and therefore they make their own selection, whereby most households resort to used plastic bags. This choice makes regular collection necessary as their limited size means they are rapidly filled up; the absence of lids increases health risks; and possibilities for waste separation are severely limited.

Cultural values and social norms turn out to be constraining factors in the contribution and use of sanitary technologies. They can, however, also be enabling, for example when existing traditions facilitate self-organization and voluntary contributions through self-reliance in the introduction of particular technological options. In general, technological improvements in sanitary facilities require horizontal and vertical linkages and lines of communication between the different social actors involved. Particular technological options should, in general, be assessed on user-acceptance, including comfort considerations, otherwise they will not provide a solution which is sustainable in the longer term. This short discussion on the relevance of householders' cultural and social norms for constructing sanitary infrastructure does not aim to convey the message that cultures and habits are fixed for all times and excluding or constraining technological innovations in urban infrastructures. What we do want to argue is that social and cultural norms are as relevant as technical prerequisites but less malleable than technical factors.

2.4.2 Neighborhoods, Local Communities and Their CBO and NGO Actors

Faced with the challenge of providing sanitation for the urban poor in East Africa many (national as well as local) authorities have considered either large-scale centralized systems or individual household-based solutions. Options at the intermediate level of the neighborhood have hardly been contemplated while they may fit the gap between pit latrines and central sewerage and allow for improvements at the neighborhood scale (Mara and Alabaster 2008). There is a need for the elaboration of concrete technical and managerial options for such alternative communal systems.

It is particularly at the neighborhood level that NGOs and CBOs can become involved in developing and managing urban environmental infrastructures. Since the 1980s, civil organizations have entered the limelight as governments throughout Africa retreated in many areas of social service delivery (Bratton 1989). The groups to be distinguished are community-based organizations (CBOs, which are small membership associations relying on limited amounts of primarily local resources), national non-governmental organizations (NGOs with small professional staffs which provide support to communities), and international non-governmental organizations (INGOs; relief and development agencies with large professional staffs and budgets and field offices in many countries). NGOs and CBOs have received increasing amounts of financial support from private and official donors. International donors have started working more via NGOs, especially since the 1980s (Fritz and Menocal 2006). The proportion of total bilateral aid channeled through (I)NGOs is increasing and individual NGOs are becoming more dependent on official development aid. Many donors and academic observers consider NGOs as more reliable channels to support community development in comparison with the official government (see Box 2.3).

NGOs are generally considered flexible and innovative, disposing of dedicated and professional staff and more willing to fulfill the obligations included in detailed project descriptions. An additional characteristic attributed to NGOs and CBOs is their essential contribution in the establishment of a thriving civil society, prepared to challenge the governmental authority when necessary. However, NGOs are not automatically more cost-effective than other sectors and the sustainability in the long run of large-scale service provision by NGOs has been questioned.[5] In case of a CBO operating as a service provider contracted by foreign donors, the relation-

Box 2.3 Kisutu Women Development Trust Fund (KIWODET)

KIWODET is formed by a group of women living in a low income area in Kinondoni municipality in Dar es Salaam. The group was formed in 1998 by a group of 20 members to collect solid waste; each member contributed 100Tshs to purchase waste bags which they distributed freely among their neighbors while offering free collection. In 1999, the Dar es Salaam city council gave KIWODET a contract to collect solid waste and the organization now serves 8,331 households. Later on they also acquired a municipal contract for street sweeping. Supported and trained by the ILO, KIWODET put in place an integrated solid waste collection and recycling system, providing employment opportunities for many young men and women.

[5] Another challenge for NGOs concerns participation and democratization, because they often fail to install such mechanisms internally while the increasing dependence on official donor funding aid may erode their legitimacy within the communities where they operate (Edwards and Hume 1996).

ship with its target groups might change into a provider-client relationship, still remaining different however from the formal relationship usually established between governments and citizens.

2.4.3 City Level Planning, Decentralization and PPP

The recently introduced *decentralization policy* in the countries of East Africa allows local populations to promote solutions that are better adapted to their problems. Decentralization has been hailed as the most appropriate vehicle for achieving good governance invoking as essential political accountability, freedom of association and participation, a reliable and equitable legal framework, bureaucratic transparency, availability of valid information and effective and efficient public sector management (Onyach-Olaa 2003). In this way, sub-national governments are better positioned vis-à-vis the local population to identify local preferences for infrastructure technologies or service quality, and facilitate local decision-making. More room may be opened up for cities to create their own solutions for sanitary infrastructures, to be better adapted to the specific local conditions and demands. The decentralization process going on in the three countries concerned here – Kenya, Tanzania and Uganda – is however not identical and varies in some important respects.

All three East African countries show some form of decentralization: political decentralization through the devolution of power to sub-national units and administrative decentralization through the creation of offices and the deployment of staff to lower levels, while fiscal decentralization is evident in the reallocation of resources. In Uganda decentralization is entrenched in the legislation and has been put into practice, but in Tanzania much of the policy reforms and the very detailed enabling legislation has as yet remained paper work primarily. In Kenya, however, decentralization is practiced to a certain extent, but has not been supported by legislation or political declarations.

In Uganda and Tanzania, staff for key social service delivery sectors (in particular primary education and health) has been transferred from central to Local Governments (LGs). The increase in numbers is very substantial; in Uganda from 65% in 1998 to 73% of total staff numbers in 2002; in Tanzania from 57% to 63%, while a further increase is foreseen for the future (Steffensen and Tidemand 2004). In Kenya, the situation differs substantially, with an increase from 12% in 1998 to only 14% in 2002 (ibid), as the LGs have a much more marginal role in service delivery. LGs (except a few of the larger municipalities) are not responsible for the key sector areas – health and education – and thus many of the functions within these sectors are carried out by agencies other than the LGs (like Ministries, Water Boards, etc.).

Public-Private Partnerships (PPPs) have been heralded in the 1990s as an attractive instrument for public policy implementation. Bringing together the best from the public and the private sector, they should facilitate the creation of cost-efficient and effective delivery services, including sanitation. Although understandings of the concept and its implementation in practice vary widely, in general governments are involved through the development, implementation and enforcement of regulation, while the private sector assumes particular responsibilities for the collection, transport and disposal of sanitary and solid waste flows. Private sector involvement may range from contracting out services to large private firms, via concessions to local entrepreneurs, to franchises to CBOs, and combinations of these.

For example, since Kenya's transfer into a multi-party system of governance in the early 1990s, the Local Government Act allows local authorities to contract out certain services to the private sector. Currently, different forms of arrangements are in place ranging from franchisees to concessions. An example is Kisumu Water and Sewerage Company (KIWASCO) managing water and sewerage in Kisumu, which is run as a private company and owned by Kisumu City Council.

Realizing the potential merits of this model in the context of Africa proved however more challenging than initially expected. In particular, the absence of a well-developed private sector, the persistent poverty among the urban poor, the lack of transparency in the operation of urban governmental institutions, and the fragmented institutional framework prevented many PPPs from becoming successful overnight.

2.4.4 National Level Policies and a Network Approach to Governance

National governmental authorities remain important as they should offer the necessary strategic guidance and holistic perspective to ensure consistency across the sanitation flow. They are under continuous pressure to deliver public services but have been unable so far to secure them. Despite good intentions at the moment of independence, most African states proved unable to fulfil their developmental role. Relevant environmental policy and legal frameworks for promoting sustainable sanitation do exist, the efforts and measures to develop concrete infrastructures and systems however often remain well-intended but hardly effectively implemented. Explanations for these 'policy gaps', as well as the concomitant recipes to address them, have changed over time.

Government failures in Africa in the 1950s were explained as the result of poverty and conceptualized as the lagging behind of economic and technological development caused by a lack of savings. Later, in the 1960s (neo-) colonialism and dependency from the 'capitalist West', were considered the main causes, but since the 1980s the attention has shifted to the quality of social institutions. The unresponsiveness of administrative systems and weak institutions in general are considered the central causes for lack of development in Africa (Kumssa and Mbeche 2004). Until the early 1990s many African countries opted, in response, for a strategy to expand and modernize the public sector to support social and economic development. Government expenditures increased from about 15% in 1960 to about 28% in 1990 of GDP (World Bank 1997). Despite this trend, most governments remained weak and did not have sufficient capacity to enforce laws and policies, even if they were enshrined in government documents. A growing number of experts pointed at the lack of attention to the role of institutions and the importance of 'good governance'. They claimed that development involves more than just adapting macro-economic policies and trimming the state bureaucracy as was promoted by the Structural Adjustment Programs (SAP) of the IMF (Kumssa and Mbeche 2004).

The discourse on 'good governance' emerged in the 1990s when donors witnessed democracies outperforming their authoritarian counterparts in economic and social development (Alence 2004; Doornbos 2001). Economic reforms were considered more likely to be sustainable and effective, if the governments imposing the transitory pain of adjustment were viewed as legitimate by the members of society. Democratic governments are better qualified to consult major social and interest groups and to involve them in the design of policies. They could – along with independent media and policy centres – do a better job in educating the public about the need for reform (Mkandawire 2006). The 'good governance' – discourse contributed to a shift from externally oriented to internally oriented conditionalities in international developmental aid as it concerned the structuring and operation of recipient countries' institutions (Doornbos 2001).[6] However, rather quickly the concept began to lose its popularity, as several countries were able to circumvent some of its prescriptive, particularly political, elements. In addition, the application of universal standards in a variety of different contexts proved complicated.[7]

Minimising the state, promoting good governance and attributing more tasks to markets and private companies did not produce the expected results either, as became clear in the early 2000s. As a consequence the role of the African state in providing urban environmental services is object of renewed discussion. Two competing views can be identified in this debate, on the one hand the neo-developmental state and on the other the network governance approach. According to the proponents of the neo-developmental state, the resources of the governmental agencies at the different levels should be strengthened and in particular their capacities to plan, implement and secure effective urban environmental infrastructure management. A network approach to governance, on the other hand, focuses on the engagement of governmental actors at different levels and of private actors (companies, NGOs, CBOs and communities) in designing and implementing urban environmental infrastructures and services.

Both views have their strong points as well as several weaknesses. Regarding the neo-developmental approach it is argued that the state in Africa has so far not been able to fulfil its responsibility as promoter of development and it is therefore extremely unlikely that this situation will fundamentally change in the near future,

---

[6] Although the 'good governance' concept and the related political and scientific debates cannot be elaborated extensively here, it is essential to note that it refers to two different aspects: the performance aspect of governance and its representational aspect (Harpham and Boateng 1997).

[7] "Introducing conditionalities often meant inserting new, specific elements into highly complex processes and situations, leading up to new complexities for which donors and recipients would henceforth bear joint responsibility" (Doornbos 2001: 102).

for example when trying to secure solid waste and sanitation services. The limited capacity, continued corruption and politicisation of the bureaucracy and the lack of adequate resources will persist and thus hamper effective interventions. According to the perspective of network governance (urban) governmental authorities are no longer pivotal in securing urban environmental infrastructures and services. This approach creates room for the effective participation of stakeholders and allows local communities to develop their own preferred approach while harmonizing their efforts with other stakeholders and with other levels of governance. Network approaches to urban environmental governance become particularly interesting when the notion of centralized infrastructures is substituted with the idea of smaller communal (infra)structures, such as those described by Mara and Alabaster (2008) and Mara et al. (2007), because they facilitate coproduction of solutions which better fit local conditions (Ostrom 1996). Network-based approaches are criticized for the lack of legitimacy, because unlike state-based regulators, whose actions are legitimized via formal democratic procedures and supported by law, non-state actors cannot rely on legal authority, nor derive legitimacy from their position in a wider legal order. For networks, legitimacy is rooted in the acceptance of their role by other stakeholders and in the multiple narratives that can be constructed in the context of multiple accountability relationships meeting divergent legitimacy claims. Community-based environmental service arrangements are not just visible entities but also manifestations of negotiated social practices, located in wider historical, economic and social contexts. Their flexibility makes such network approaches more suitable to respond to specific local conditions and public demands, allowing a better fit between technological options, their management and the prevailing societal circumstances.

2.5 Conclusions

The absence of large-scale environmental infrastructures in many East African cities, the fast growing urban populations and the growing concerns about environmental and health impacts resulting from the use of traditional pit-latrines create fertile conditions for introducing innovative solutions. The Modernized Mixtures approach can provide a coherent perspective for transition processes towards establishing more sustainable sanitary systems and services that are also accessible for the poor. This approach does not signify the identification of one ultimate solution for sanitary problems but is aimed at the identification of multiple pathways towards the creation of improved sanitary infrastructures in the different cities in East Africa and beyond.

The Modernized Mixtures approach presents an integrative perspective on sanitation in informal settlements in the cities of East Africa. It stands for combining technological and social dimensions, integrating the management of sanitation and waste flows from generation to final discharge, and linking the different technological and social levels of scale involved in the provision of environmental infrastructures and services. This chapter has further elaborated the MM-approach by pointing in particular to the social, cultural and political dimensions of the design, management and use of urban environmental infrastructures at household, community, city and national levels. The different options at the distinguished levels of scale allow a modular approach to the choices of technology and management systems. Nevertheless several key challenges remain to be addressed.

Introducing the Modernized Mixtures approach is new, uncharted territory in infrastructure development and therefore involves some risk-taking and requires long-term commitment from relevant actors, including foreign donors. Consistent political engagement is required to develop innovative technologies which consider sanitation and waste as an integrated material flow and promote prevention, reuse and recycling options instead of simply discharging. Long term commitment is however not assured in a simple manner as political priorities may change while the continuing dependency on external donors in financing infrastructures makes consistent policies also dependent on their shifting priorities. Strengthening the social, scientific and governance networks that are engaged with the design, development, maintenance and use of environmental infrastructures at different levels of scale seems to be the best feasible way forward.

References

Alence, R. (2004). Political institutions and developmental governance in Sub-Saharan Africa. *Journal of Modern African Studies, 42*, 163–187.

Boadi, K., Kuitunen, M., Raheem, K., & Hanninen, K. (2005). Urbanisation without development: environmental and health implications in African cities. *Environment, Development and Sustainability, 7*, 465–500.

Bratton, M. (1989). The politics of government-NGO relations in Africa. *World Development, 17*(4), 569–587.

Doornbos, M. (2001). 'Good governance': the rise and decline of a political metaphor. *Journal of Development Studies, 37*, 93–108.

Edwards, M., & Hume, D. (1996). Too close for comfort? The impact of official aid on nongovernmental organizations. *World Development, 24*(6), 961–973.

Fritz, V., & Rocha Menocal, A. (2006). *(Re)building developmental states: from theory to practice*. London: Overseas Development Institute.

Geels, F. W. (2004). Understanding system innovations: a critical literature review and a conceptual synthesis. In B. Elzen, F. W. Geels & K. Green (Eds.), *System innovation and the transition to sustainability; theory, evidence and policy* (pp. 19–47). Cheltenham: Edward Elgar.

Geels, F. W. (2005). *Technological transitions and system innovations: a co-evolutionary and socio-technical analysis*. Cheltenham/Northampton, MA: Edward Elgar.

Geels, F. W. (2006). Multi-level perspective on system innovation: relevance for industrial transformation. In X. Olshoorn, & A. J. Wieczorek (Eds.), *Understanding industrial transformation: Views from different disciplines* (pp. 163–186). Dordrecht, The Netherlands: Springer.

Geels, F. W., Elzen, B., & Green, K. (2004). General introduction: system innovation and transitions to sustainability. In B. Elzen, F. W. Geels & K. Green (Eds.), *System innovation and the transition to sustainability; theory, evidence and policy* (pp. 1–16). Cheltenham: Edward Elgar.

Giddens, A. (1984). *The constitution of society. Outline of the theory of structuration*. Cambridge: Polity Press.

Guy, S., & Marvin, S. (1996). Transforming urban infrastructure provision–the emerging logic of demand side management. *Policy Studies, 17*, 137–147.

Harpham, T., & Boateng, K. A. (1997). Urban governance in relation to the operation of urban services in developing countries. *Habitat International, 21*, 65–77.

Hegger, D. (2007). Greening sanitary systems, an end-user perspective. Ph.D. thesis, Wageningen University, Wageningen.

Hukka, J. J., & Katko, T. S. (2003). *Water privatisation revisited: panacea or pancake?* Delft: IRC, International Water and Sanitation Centre.

Kemp, R., Schot, J., & Hoogma, R. (1998). Regime shifts to sustainability through processes of niche formation. The approach of strategic niche management. *Technology Analysis and Strategic Management, 10*, 175–195.

Kombe, W. J. (2005). Land use dynamics in peri-urban areas and their implications on the urban growth and form: the case of Dar es Salaam, Tanzania. *Habitat International, 29*, 113–135.

Kumssa, A., & Mbeche, I. M. (2004). The role of institutions in the development process of African countries. *International Journal of Social Economics, 31*, 840–854.

Kyessi, A. G. (2005). Community-based urban water management in fringe neighbourhoods: the case of Dar es Salaam, Tanzania. *Habitat International, 29*, 1–25.

Lens, P., Lettinga, G., & Valero, M. (2003). Environmental protection technologies for sustainable development. In J. Marsalek, D. Sztruhar, M. Giulianelli & B. Urbonas (Eds.), *Enhancing urban environment by environmental upgrading and restoration* (pp. 321–329). Dordrecht/Boston, MA/London: Kluwer.

Lens, P., Zeeman, G., & Lettinga, G. (Eds.). (2001). *Decentralised sanitation and reuse: concepts, systems and implementation*. London: IWA Publishing.

Mara, D., & Alabaster, G. (2008). A new paradigm for low-cost urban water supplies and sanitation in developing countries. *Water Policy, 10*, 119–129.

Mara, D., Drangert, J.-O., Anh, N., Tonderski, A., Gulyas, H., & Tonderski, K. (2007). Selection of sustainable sanitation arrangements. *Water Policy, 9*, 305–318.

Mels, A., Kujawa, K., Wilsenach, J., Palsma, B., Zeeman, G., & Loosdrecht, M. v. (2005). Afvalwaterketen ontketend. Utrecht: Stowa.

Mistra. (2002). *The urban water Mistra programme, annual report 2002*. Stockholm: MISTRA.

Mkandawire, T. (2006). *Disempowering new democracies and the persistence of poverty. Democracy, governance and human rights programme paper 21*. Geneva: UNRISD.

Mol, A. P. J., & Spaargaren, G. (2006). Towards a sociology of environmental flows: a new agenda for twenty-first-century environmental sociology. In G. Spaargaren, A. P. J. Mol & F. H. Buttel (Eds.), *Governing environmental flows; global challenges to social theory* (pp. 39–82). Cambridge, MA: MIT Press.

Mubeezi, R. (2007). *Decentralization of environmental health services in Uganda: A case study of Mpigi and Mukono districts*. M.Sc. thesis, Wageningen University, Wageningen.

Nilsson, D. (2006). A heritage of unsustainability? Reviewing the origin of the large-scale water and sanitation system in Kampala, Uganda. *Environment and Urbanization, 18*(2), 369–385.

Nkurunziza, E. (2007). Informal mechanisms for accessing and securing urban land rights: the case of Kampala. *Environment and Urbanization, 19*, 509–526.

Onyach-Olaa, M. (2003). The challenges of implementing decentralisation: recent experiences in Uganda. *Public Administration and Development, 23*, 105–113.

Ostrom, E. (1996). Crossing the great divide: coproduction, synergy, and development. *World Development, 24*, 1073–1087.

Otterpohl, R., Grottker, M., & Lange, J. (1997). Sustainable water and waste management in urban areas. *Water Science and Technology, 35*, 121–133.

Sano, J. C. (2007). *Urban environmental infrastructure in Kigali city, Rwanda. Challenges and opportunities for modernised decentralised sanitation systems in poor neighbourhoods*. M.Sc. thesis, Wageningen University, Wageningen.

Satterthwaite, D. (2004). *The under-estimation of urban poverty in low and middle-income nations*. London: IIED.
Scheelbeek, P. (2006). *Urban environmental infrastructure around Lake Victoria: Challenges and opportunities of decentralized sanitation systems for the urban poor.* M.Sc. thesis, Wageningen University, Wageningen.
Schumacher, E. F. (1973). *Small is beautiful: a study of economics as if people mattered.* London: Blond Briggs.
Schwartz-Cowan, R. (1987). The consumption junction: a proposal for research strategies on the sociology of technology. In W. E. Bijker, T. P. Hughes & T. J. Pinch (Eds.), *The social construction of technological systems: new directions in the sociology and history of technology* (pp. 261–280). London: Guilford Press.
Seppälä, O. T., Rajala, R. P., & Katko, T. S. (2004). Consumer responsive water and sanitation services. *JAWWA, 96*, 83–93.
Spaargaren, G., Oosterveer, P., Van Buuren, J., & Mol, A. P. J. (2006). *Mixed modernities: towards viable urban environmental infrastructure development in East Africa.* Wageningen: Wageningen University.
Steffensen, J., & Tidemand, P. (2004). *Decentralisation in East Africa, a comparative study of Tanzania, Uganda and Kenya, syntheses report.* Taastrup: NCG, for the World Bank.
UNDP-International Poverty Centre. (2005). *In Focus 7: Poverty and the City*. Brasilia: UNDPO-International Poverty Centre.
UN-Habitat. (2006). *State of the world's cities 2006/7. The millennium development goals and urban sustainability: 30 years of shaping the habitat agenda.* Nairobi: UN-Habitat.
Van Vliet, B. (2006). The sustainable transformation of sanitation. In J. P. Voss, D. Bauknecht & R. Kemp (Eds.), *Reflexive governance for sustainable development* (pp. 337–354). Cheltenham: Edward Elgar.
Walker, W. (2000). Entrapment in large technology systems: institutional commitment and power relations. *Research Policy, 29*, 833–846.
World Bank. (1997). *World Bank development report 1997: The state in a changing world.* New York: Oxford University Press.

Chapter 3
Sense and Sanitation

Bas van Vliet and Gert Spaargaren

Abstract Historically, sanitation infrastructures have been designed to do away with sensory experiences. As in the present phase of modernity the senses are assigned a crucial role in the perception of risks, a paradigm shift has emerged in the infrastructural provision of energy, water and waste services. This has led to a partial re-localization and resensitization of services. Present systems are designed to make the invisible visible again. This chapter analyzes what these tendencies mean for waste water and sanitation service provision. It outlines the paradigm shifts being made in infrastructural provision and its consequences for the senses, using case studies of sanitation innovation in Europe to illustrate new dynamics in the display and perception of sanitation infrastructures. Based on a theoretical discussion of sensitization of infrastructural service provisions, a framework is presented for analyzing the possible relationships between senses and sanitation.

3.1 Introduction

With his concept of World Risk Society Ulrich Beck (2007) indicates the fact that risks – and environmental risks in particular – have become key factors in determining the nature of modern societies, their cultural and political institutions and their ways of perceiving risks. Contrary to the earlier phase of industrial development, the present phase of 'reflexive modernization' assigns an important role to scientific experts and engineers for interpreting and framing risks, since most of the risks have become inaccessible for lay people. For assessing the dangers of modern risks like climate change, BSE in food chains or radiation, our senses are no longer a reliable instrument. We are forced to rely upon and trust the technologies and assessments of experts who mostly operate out of sight and at a distance from our ordinary lives

---

B. van Vliet (✉) and G. Spaargaren
Environmental Policy Group, Wageningen University, Hollandseweg 1, 6706 KN, Wageningen, The Netherlands
e-mail: bas.vanvliet@wur.nl; gert.spaargaren@wur.nl

B. van Vliet et al. (eds.), *Social Perspectives on the Sanitation Challenge*,
DOI 10.1007/978-90-481-3721-3_3,

and daily experiences. However, since we are aware of the fact that scientists are no longer in the position of providing undisputed, solid and single best answers for dealing with risks, in the World Risk Society there is a need to re-invent risk politics and to restore the role of the senses.

Risk politics in reflexive modernity includes the active involvement and commitment of citizen-consumers, their (health) worries, their (consumer) preferences and their interests for sustainability. One of the crucial elements and in fact the pre-condition for organizing the active involvement of citizen-consumers in risk-politics is 'bringing the systems of production and consumption back' into the life-world. For only if people have a vision, a direct sensory relationship to the processes which are behind the risks they face, a mature form of risk policy can emerge. Risk politics must be (re)connected to our ways of 'dwelling the place' (Urry 2000). In this connection, sensory experiences and visible awareness of the risk-producing systems gain a direct political significance. Next to the food system the infrastructures for the provision of water, energy and waste services are among the most intricate and intimate systems organizing our lifes. As such, sanitary systems are the perfect case for illustrating the rediscovering of sight, smell and the senses in general when dealing with sanitary infrastructures and the risks they represent.

Our understanding of the world of sanitation and waste water management is dominated by an engineering language of material flows, nutrients and infrastructures. Drawing upon this, environmental engineers and health workers in sanitary systems try to implement the best technologies in terms of nutrient reuse, water saving, biogas production or pathogens removal. In doing so, they risk to overlook the intimacy and everyday life aspects of sanitation practices. Contrary to the language of flows, nutrients and pathogens, end-users considering their sanitation practices may think in 'cultural' terms of (risk) experience, comfort and cleanliness instead.

Sanitation systems belong to what we have called 'infrastructures of consumption' (Van Vliet et al. 2005). These are the infrastructures through which energy, water and waste services are provided to domestic consumers. Taken together, these service provisions comprise a major part of domestic consumers' environmental resource use.[1] During the twentieth century, infrastructures of consumption have been rolled out in the urban centres of the world, and in ever increasing scales. The current division in responsibilities between citizen-consumers on the one hand and providers within infrastructural systems on the other can be characterized as one of lay people versus experts. Citizen-consumers are expected to be only using and paying for the services while providers are responsible for standard setting, supply, maintenance, monitoring and billing. This division dates back to the period when infrastructures of consumption were being up-scaled and connected to regional and national grids. Since then the systems have turned into 'abstract' systems that are largely invisible to their end-users.

[1]Some figures from Milieu en Natuur Compendium (2008): Household consumers in the Netherlands were responsible for about 750 million cubic meters of drinking water production, the emission (via sewerages) of 76% of all Nitrogen to open waters; they are directly responsible for the production of 15% of all solid waste and they directly consume 19% of total energy production (excluding fuels for traffic). (www.milieuennatuurcompendium.nl)

And since invisible also means out of sight and unknown, a divide has grown between the experts in systems of provision and their domestic end-users.

To overcome this divide attempts have been made since the 1980s to make water and energy infrastructures visible or tangible again to their end-users. It was assumed that increased visibility of water and energy resources and their infrastructures would lead to a better understanding of why and how these systems are designed, operated and maintained. From this increased understanding, a more rational resource use from the side of the end-user was expected to emerge. Visible water drainage in residential neighbourhoods and the (conspicuous) integration of solar panel systems in the built environment may serve as examples. While in general the re-sensitizing of infrastructures of consumption are judged to be positive phenomena, sanitation infrastructures present a different case in this respect.

3.1.1 Outline of the Argument

We firstly explore the role that sensory experiences can play and historically have played in the development of waste water infrastructures. From this analysis we try to assess what roles sensory experiences can play within contemporary innovations in sanitation. In the next section we sketch the gap that has emerged between 'abstract' systems and their experts on the one hand, and consumer practices of energy and (waste)water handling on the other. We stress the implications of the paradigmatic changes for the sensory experiences related to utility services in general and to waste water and sanitation infrastructures in particular. In section three we give a theoretical explanation of what 'making the invisible visible' (Shove 1997) actually entails and what exactly is made visible, or sensible and for what reasons. The following section presents a number of typical experiences with innovation in sanitation, obtained from research projects[2] conducted mainly in the Netherlands. We conclude with remarks on the (non)sense of appealing to the senses in the design and use of sanitation infrastructures.

3.2 Paradigmatic Changes in Infrastructures of Consumption

How did sanitation networks and other infrastructures of consumption become abstract systems? For householders, infrastructural systems started to become 'abstract' as soon as domestic daily chores were gradually taken over by new infrastructural services. In her classical study on the industrialization of American households,

[2] DOMUS: Domestic Consumption and Utility Services, EU Framework Four funded project (1997–2000) on environmental innovation in water, energy and waste sectors in the Netherlands, United Kingdom and Sweden (Chappells et al. 2000). DESAR: Decentralized Sanitation and Reuse: National funded (EET) project (2001–2007) on decentralized sanitation and social opportunities and risks (Hegger et al. 2008).

Ruth Schwartz Cowan (1983) illustrates how domestic practices of cooking, laundering and cleaning the house could become fundamentally restructured because of the rolling out of water works, electricity, gas and waste collection systems which 'liberated' consumers from collecting and managing basic resources like water, waste and energy. With key domestic routines changing in character, also the sensory experiences of water, energy and waste took on different dynamics.

Once the major networks were developed,[3] the most common way of infrastructure service provision was by establishing state or municipality owned companies to service connected consumers. To safeguard standardized technology and procedures and to benefit from economies of scale, infrastructure systems often built up hierarchic organization structures with command-and-control modes of regulation. The logic of network management in such public modes of provision is predominantly supply-oriented. Expansion of utility networks is connected with the drive to improve national economic performance, public health and the quality of life. Levels of energy consumption, connection to water, sanitation and waste networks become surrogate indicators for levels of national economic performance. Guided by this logic, the specific needs of individual cities and consumers have little impact on the process of network provision and management (Graham and Marvin 1995). Not only systems have become abstract to consumers, also consumers have become abstract to system managers. In this phase of universal infrastructure service provision, householders are termed 'connections': what happens behind the meter does not count.

In OECD countries the supply-driven logic in energy, water en waste provision was dominant until about the 1980s (Graham and Marvin 1995; Juuti and Katko 2005). The first seeds of change were initiated by (projected) resource scarcities, or unacceptable environmental damage caused by ever increasing scales of provision of services. The oil crisis in the beginning of the 1970s resulted in strategies of demand side management, while also energy conservation became part of the service packages energy companies offered to their consumers. Although demand side management is primarily driven by providers rather than by consumers (see Van Vliet et al. 2005; Chappells et al. 2000), it does bring consumer issues more to the foreground. For the first time since the major networks had been rolled out, the 'taken for grantedness' of infrastructural service provision had been challenged. Consumers were addressed to consider the timing and volumes of energy and water consumption, and to monitor water and energy flows. The first steps were made in making the invisible visible again.

The real paradigm-shift in the logic of water, energy and waste provision however only happened during the 1980s. By that time state-owned utility companies were increasingly accused of being over-staffed, inefficient, stagnant and not responding to market- and consumer-demands. The resulting privatization of formerly state-owned utility companies and the liberalization of water, energy and waste service provision have shaped the public discourse on utility provision all over the world

[3] Fifty percent of American Cities had waterworks already in 1830 (Melosi 2000: 74). In most European countries municipal Waterworks emerged around 1880 (Juuti and Katko 2005).

(Estache 1995; Kessides 2005). At the same time, the environmental performance of utility systems started to become addressed in international environmental policy making. As a result of these tendencies, formerly uniform services and networks have become unbundled and splintered into special niche markets (green electricity) and smaller units (house-on-site grey water recycling), allowing for differentiation in service provision. One of the many implications of this paradigmatic shift in infrastructural service provision is that consumer roles in infrastructure service provision have shifted from being 'connections' or captive consumers towards being customers, co-providers or citizen-consumers (Van Vliet 2003). A renewed emphasis on the role of consumers in infrastructural service provision has come along with attempts to make systems of provision transparent and resource use 'visible' again to their end users. So while infrastructural service provision has become embedded in globalized networks of liberalized energy, water and waste markets, at the same time the servicing of end-users tends towards regionalization and (lifestyle-) differentiation: more visible, locally accountable and accessible, transparent systems with down-scaled forms of supply while meeting the demands of particular groups of end-users.

3.3 Sanitation as a Special Case

This general pattern of two major paradigmatic changes in infrastructures of consumption in the nineteenth and twentieth century largely applies for systems that deal with domestic waste water as well. Also in sanitation from the 1980s onward we observe a trend towards retaining and reusing waste water flows, and gradually paying more attention to the role of end-users in the design and functioning of these systems. The consumer starts being represented in Water Boards and efforts have been made to increase transparency and accountability of the waste water systems and their managers. In stead of only one – centralized – solution to the sanitation challenge also decentralized solutions are being applied in pilot projects. There is however one aspect of sanitation which makes this utility service an exception to the main trends. Where energy and water-flows are being re-introduced into everyday life and made visible, tangible, trustworthy and enjoyable for its users, 'bringing back the senses' in the field of waste water and sanitation is highly problematic. This is because it seems detrimental to the design principles of sanitation systems as we know them and because it would clash with some of the deeply rooted standards and cultural believes that have been built around sanitation and hygiene over the past century. We will briefly discuss this special status and position of sanitation infrastructures of consumption below.

Unlike any other urban infrastructural system, sanitation systems have been designed just to avoid contact with humans and to diminish the sensory experiences of sight, touch and smell. Before modern sewer systems were founded, the 'miasmatic' theory prevailed, in which "disease was understood to arise from putrefying organic wastes, bad smells (miasmas), and sewer gases" (Melosi 2000:47). It linked the notion of bad smell to that of danger and disease. The nose was by far the most important instrument to assess environmental risks. For humans in the nineteenth century the greatest risks were those that could be smelled, in particular those posed by decaying organic matter (Hegger 2007:75). After the miasma ideas had been replaced with epidemiologic notions of direct contamination with bacteria (in particular by Pasteur), not only the avoidance of smell of human waste, also that of touch and ingestion were added to the design principles of water and sewer works.

Since water and sewer works have been designed to take away or avoid sensory experiences, 'bringing back the senses' in waste water services runs counter to some of the very design principles of modern sanitation systems. Besides, starting to use the nose and the eyes again would clash with deeply rooted social-cultural norms and practices of the avoidance of smell and contamination with pathogens. It would provoke a reaction now commonly called the 'yuck factor'. Russell and Hampton (2006) refer to the yuck factor as "an emotional response to the idea of reusing water derived from sewage" (ibid p. 218), which can easily be extended to any sensory experience from end-users related to sewerage. Not surprisingly, the 'yuck factor' is embraced by local authorities and water managers as an easy argument not to furthermore play and experiment with local water reuse and on-site waste water treatment.

So waste water infrastructures in some respects can indeed be regarded the exception to the rule when it comes to re-sensitization in infrastructures of consumption. This does not imply however that sanitation should remain 'senseless' at all times and in all respects. To specify which aspects could be brought back in and for what reasons, we need to take a closer look at the theory behind (re)sensitizing infrastructures.

3.4 The Theory Behind (re)Sensitization

When Ulrich Beck published his influential book on the 'Risk Society' (firstly in 1986) the Chernobyl disaster served as the tangible example to illustrate the central thesis of his study. While it was clear to everyone that a major disaster had happened, at the same time it proved impossible for lay-actors in Europe to assess the immediate threats of the nuclear melt-down for their everyday lives. Could you go out door? Would it be wise for water companies to temporarily use groundwater instead of surface water as a source for drinking water?

The main reason for not being able to determine an adequate response to the Chernobyl disaster was the fact that people could not rely on their sensory equipment (their noses, ears and eyes) to detect and assess the dangers. While in earlier times sensory experience was a reliable tool to assess environmental risks, in the mid-1980s environmental problems had increasingly become invisible, abstract problems. As a result, sensory experiences no longer function as a reliable compass to guide the risk (avoidance) behaviour of lay-people.

For making up their minds, people have come to rely on scientific experts who assess the environmental dangers with the help of scientific methods instead of using (only) the sense organs of the human body. Experts and their abstract systems are not just the ones providing us with products like energy, water, food or clothes, they are also our major source of information about the environmental risks and side effects which come along with the industrial organization of these product-flows.

At the same time, however, most lay-people are by now aware of the uncertainties, limitations and contradictory claims of modern scientific works and the expert systems they help organize. Science and technology mobilized for sustainable development tend to fall victim to intense societal debates and are confronted with counter-claims with respect to the environmental and social (side) effects of the technologies put forward. As a result, people are dependent from and at the same time more insecure about the experts, the technologies and the abstract systems which are involved in their consumption routines. Because being 'trusted' by their users is now of key importance to the providers of services, their core-business is now to re-establish their relationships with end-users.

Our short excursion into Becks' analysis on changing relations between expert-systems and their end-users teaches us that the popular concept of 'transparency' refers to several dimensions of the relationship between providers and end-users. Next to the enhanced visibility of the system and its experts in a factual sense, also the (environmental) performance of the system has to be displayed in public, before the eyes of the beholders. These undertakings from the side of abstract systems are aimed at maintaining adequate levels of trust from the side of lay people. By making themselves visible to and by actively seeking contact with lay actors, the experts at so called 'access points'[4] inform lay-people about the quality and reliability of their products and services. In this way, utility companies seek to deserve their licence to produce from end-users. In the sections to follow we will explore the reasons behind (re)sensitization of utility services and infrastructures as well as the methods to implement it.

3.5 Reasons for Enhancing the Visibility of Utility Provision and Consumption

There are at least three different reasons for making utility provision more visible to end-users. First, visible feed-back can result in consumers looking for ways to save money and resources and hence to increase the eco-rationality of their behaviours. Second, consumers can use (green) utility products and technologies as symbolic media to express their (green) lifestyles. Third, becoming involved in collective experiments of green utility provision and consumption can be used to demonstrate the need for sustainable development to fellow citizens and competing companies in the wider society.

[4] Access points are described by Anthony Giddens as the points where lay-people get in contact with representatives of the expert systems, like the butchery for the food chain or the doctors' room for the medical system (Giddens 1990).

1. *(Eco)rationalization through feedback.* Most campaigns in energy or water saving start with monitoring, metering and data-logging at household levels (Van den Burg 2003). In essence, monitoring means: making the so far invisible practices of resource use visible. It is assumed that knowing one's energy or water consumption – and the economic and environmental costs attached to it – will lead to a more rational or efficient use of these resources. Although monitoring devices and feedback schemes are mostly targeted at visualizing the behaviour of consumers at the downstream side of production-consumption chains, it is not restricted to end-users only. Companies can organize visible feed back on their own behaviours as well. By showing monitoring data (through specified bills, energy-mirrors attached to buildings etc.) to customers and the general public, utility companies disclose their own environmental performance. Enhancing the eco-efficiency of the performance of both consumers and providers can be done without too much politics involved, so it is argued. The most important challenge is to get the new information flows, products and services accepted and used by consumers. Information campaigns, labels, monitoring technologies and economic incentives are the main ingredients of this strategy.
2. *Using utility objects and infrastructures to play the symbolic game of (sustainable) consumption.* During the past 10 years, the limits to the (eco) rationalization discourse have become widely recognized. A narrow focus on just the economic and technological dimension of service provision may result in consumers using environmental innovations improperly or not using them at all. People consume goods and services not just for instrumental and functional reasons but also to meaningfully relate to their friends, neighbours and colleagues. Goods and services and the consumption practices they help organize are carriers of meanings, signals to the outside world about who we are and how we view the world. Hence, new, more sustainable technologies and infrastructures for energy, water and waste services should also be judged with respect to their abilities to help organize this game of display. Sensory experiences do play their part here. Consumers want power-showers and Jacuzzis to 'feel and enjoy' water and energy related consumption practices. These games of display are not limited to standard products and services, they also pertain to more sustainable products, technologies and services. Providers of domestic environmental services are aware of this symbolic dimension. They have come to present themselves as 'wellness -providers'. They have been diversifying their products and services in order to serve different lifestyle-groups of consumers (Hegger et al. 2008), including different variants of 'green' consumers. With the broadening of the rationale behind energy, water and waste service provisioning, both providers and consumers develop new roles and identities. They no longer uphold the captive, functional relationships that were so characteristic of the period of public provisioning. Along with the overall differentiation of services and relations also green alternatives increase the number of options available for display: water-saving toilets and shower heads, eco-dishwashers, solar boiler systems have become objects suitable for playing the consumption games in ways similar to the use of cars, couches or paintings. We could summarize this trend of enhancing the visibility, joyousness

and usability of utility infrastructures and services by assessing that the former *uniform consumption practices* organized around mostly hidden utility goods and services have become *diversified practices of consumption* enabled by re-sensitizing utility services (services made more visible, tangible, enjoyable).

3. *Demonstrating organized commitment for sustainable utility provision and consumption.* In our third category the emphasis is on shared, collective projects which are organized and displayed because they help envision a more sustainable future for society at large. Companies, neighbourhoods, cities, even countries become involved in experiments with renewable energy, innovative water and waste projects to experiment with the new, more sustainable products and infrastructures of the future. Windmills, solar panels, trees, waterponds have by now become crucial elements in the spatial design of cities and neighbourhoods, with urban planners and architects using these environmental technologies to enhance the quality of life for citizen-consumers (Kristinsson and Luising 2001). Experiments with climate neutral housing or closed-loop water systems have a strategic relevance for the (public and private) providers involved, since they expect new markets, export-possibilities or new votes from it. Although many pilot projects in sustainable utility provision and consumption are developed from an expert or provider point of view, this is not necessarily or exclusively the case. Next to provider driven, top-down organized projects there have been developed bottom-up, citizen-consumer initiated projects, as Dries Hegger (2007) has shown. Groups of citizen-consumers act as fore-runners, experimenting with sustainable housing at larger scale, try to set new standards to be adopted by mainstream providers later on. This 'demonstration' dimension bears some overlap with the second category because sustainable housing projects are often meant to display the green lifestyles of their inhabitants as well. The difference with the second category is the orientation of the projects: they are set-up as demonstration projects to show at the societal level that a more sustainable society is possible. And rather than individuals playing the symbolic games of consumption this third category refers to organized groups of citizens demonstrating their political commitment to a sustainable future. Visualization becomes demonstration for a better future.

3.6 Methods of (re)Sensitizing Flows, Infrastructures and Practices

In this section we explore the different ways in which sensitization can be organized. The first important question is what exactly will become sensitized. Building upon the analysis of provision and consumption in terms of material flows which travel through networks and scapes (Spaargaren et al. 2006) we argue that sensitization strategies can be targeted at (1) the material flows themselves, (2) the infrastructures, scapes and networks along which the flows travel, and (3) the domestic practices of consumption made possible by and utilizing the material flows.

1. *(Re)sensitizing material flows.* Material flows like water, waste, waste water and energy offer ample opportunities for both direct and indirect re-sensitization. In direct terms, flows can be made visual and tangible again by opening up formerly hidden infrastructure of (waste) water or energy, as it is being done in eco-neighborhoods. Here we see rainwater or grey water collected in open ditches and reed bed filters instead of these flows being collected in underground mixed sewer pipes. In an indirect form, the flows are increasingly being visualized by innovative metering and billing schemes and systems. Meters and displays visualize physical flows of water and energy in cubic meters and kilowatt hours or in monetary units. Billing devices visualize the (reverse) financial flows attached to resource use; show 3-year average flow consumption; compare individual consumption levels to average consumption in the neighborhood, etc.
2. *Visualizing the infrastructures, networks and scapes through which flows travel.* With the decentralization of production and differentiation of services the infrastructure hardware re-enters the domains of households and neighborhoods. Visible solar panels partly replace the abstract central electricity production units, just as decentralized water treatment systems do for the central sewer and waste water treatment systems. In some cases pipes and cables are deliberately brought back and visualized in public space for aesthetic reasons. Especially in the water sector, pieces of infrastructure are highlighted with the help of information panels informing to tell the public about the functioning of the drinking water or sanitation system.
3. *(Re)sensitizing the practices of consumption which flows help to organize.* Forms of conspicuous consumption were until recently reserved for consumption practices like housing and furnishing, car driving or, choosing one's holiday destination, but increasingly the practices of consumption around water, energy and waste services become conspicuous forms of consumption as well. Striking examples may be cooking in an outdoor kitchen, or bathing in Jacuzzis. Both kind of practices show the increasing role of display and the senses (visibility, sound, smell and bodily feelings of wellness). Although less hilarious, other visualizations are green labels for using eco-power or certificates for Climate Compensation of air traveling.

3.7 Sanitation as a Special Case

Can waste water and sanitation services help shape the symbolic en sensory game in similar ways as cars or prestigious objects do? Based on our argumentation for seeing sanitation as a special case in 'bringing back the senses', we are inclined to think that strategies for visualization and (re)sensitizing are much more restricted when compared with cars, clothes or even drinking water or energy. And if things are being sensitized in waste water systems, we expect the methods to refer to the indirect visualization of the material flows and to their infrastructures, rather than to the sanitary flows and the related consumption practices themselves. With respect to the reasons behind sensitizing strategies, we expect that most attempts will be driven by (1) the (eco)rationalization or resource use and the need to (3) demonstrate organized commitment to sustainable development, while we hardly expect to find cases of (2) the use of (diversified) sanitation practices for cultural display. To explore these hypotheses in more detail and to determine the chances for future attempts at (re)sensitizing of sanitation provision and consumption, we discuss in the next section our empirical cases.

3.8 Cases of Innovation in Sanitation

With our newly developed toolbox of reasons behind and methods of sensitization we can now revisit a number of key cases of sanitation innovation that we have been studying over the last 10 years in the Netherlands and a number of other European countries (Hegger et al. 2008; Chappells et al. 2000). Our case studies in energy, waste and water infrastructure innovations were done to find evidence for the second paradigm shift in infrastructure provision and the changing relations between consumers and providers. Also in water and sanitation infrastructures we have therefore studied the innovations that are exemplary for a change in scale of service provision, and/or differentiation of services. In the following we discuss the sensory aspects of the implementation of:

1. *Neighbourhood on-site vacuum toilet systems and anaerobic digestion.* The innovation here is the combination of technologies applied and the application at residential neighbourhood level. Anaerobic treatment has so far been applied mostly in industrial settings, while the transport of waste (water) through vacuum pipes is only known from applications in trains and aircrafts. The scale of application is new: a small vacuum network leads to on-site treatment of waste replacing huge sewer pipes and centralized waste water treatment plants (Hegger 2007). A first implementation in Wageningen was never realized due to reasons that are linked to the dynamics of sensory experiences as we will show below.
2. *Composting toilets and the Non-Olet.* Composting toilets do not need connections to piped water, nor to sewer systems. The toilet seat is mounted on a storage tank in which faeces and urine, together with some additional organic material is composted. The resulting compost can be used for gardening. Also the Non-Olet is a stand-alone toilet, but with a much smaller storage tank. The waste is to be disposed of in the organic waste bin that is centrally collected (see www.nonolet.nl). We have studied the experimentation with composting toilets in Utrecht (Chappells et al. 2000; Van Vliet 1995) and the diffusion of the Non-Olet in the Netherlands (Van Vliet, B. 2006 and Van Vliet, J. 2006).
3. *Household water projects.* During the 1990s, a number of Dutch water companies have experimented with supplying households a minor quality of water through a separate piped water system (Van Vliet 2003). This so-called household

water was meant to fill washing machines, to flush toilets, and to water the garden. The experiments were stopped, not because consumers did not accept the system, but because of misconnections being made between the pipes of drinking water and household water supply in at least two projects. After this the Ministry of Environment forbade in 2003 any further experimentation with dual water supply systems (Van Vliet et al. 2005).

4. *Grey water treatment through reed-bed filters.* Reed bed filters have been built in a number of eco-neighbourhoods in the Netherlands since the early 1990s. These filters treat waste water from washing machines, kitchen sinks, baths and showers up to a level that it can be discharged to surrounding surface waters. Reed-bed filters are typically located in and being part of the spatial plan of a neighbourhood. We have studied projects with reed bed filters in Culemborg, Groningen (Hegger et al. 2008), Arnhem, and Utrecht (Chappells et al. 2000).

In most cases consumers, providers, experts or regulators were not deliberately thinking about what the senses might do to the success or failure of their projects. Hence, applying our newly defined reasons and methods of sensitization to these cases is an exercise done in hindsight only. We can now explore what is actually made 'visible' in these projects and how this has been done, intentionally or not. In addition, we explore the possible reasons behind sensitization.

Starting with the material base – the *infrastructures* – it is clear that all the innovation projects make water and infrastructures more visible and tangible as compared to conventional, hidden infrastructures of water pipes and sewers. Most strikingly however, this counts for the reed-bed filters and household water projects. The reed beds encroach substantive parts of public space in the neighbourhoods where they are installed. There are more than only the obvious reasons behind such high visibility of reed-bed filters. Both to consumers as well as to the outside world they have become signals of sustainable lifestyles and neighbourhoods, comparable to the role solar panels have on the roofs of sustainable houses.

Also household water production required local water pumping and treatment facilities that were much more visible to their users as compared to conventional drinking water supply. The new pipes for household water were made visible throughout the system by a turquoise painting, both for practical and for symbolic reasons. Rational resource use (flushing toilets with low-quality water) was the main message that was communicated to users and society at large. On a symbolic level, the coloured pipes were a constant reminder to consumers that they were not spilling drinking water for uses like toilet flushing or gardening. Also the water companies had a stake in showing the infrastructure to the outside world as by the mid-1990s they were still prone to possible privatization[5] and eager to show their innovative capacities to the outside world.

The *flows* have been sensitized especially in the case of compost toilets and Non-Olet systems. In fact the storylines around composting toilets is one of 'closing

[5] Only in 2003, National Parliament approved a new Law securing that drinking water supply companies will remain public entities.

the loop' and human-natural cycles of nutrient flows. Also the stories of (failures of) maintenance consider the management of flows, rather than the (rather simple) infrastructure itself. Nuisance of smell and sight of the composting material, occasional nuisance of flies, and hardship of draining the urine and digging out the composting tank are all well reported in an evaluation of a failed composting toilet system project in Utrecht, the Netherlands (Post 2000; Chappells et al. 2000). Motives for sensitization as expressed by users and compost toilet builders range from rational resource use (water saving, nutrient recycling), symbolic representations (being part of natural cycles and independent from large technical networks) to displaying to the outside world that composting toilets are in fact an example for future sustainable sanitation (Huizing 1993; Van Vliet 1995).

Due to failures in proper management and maintenance of the reed bed filters in Groningen, occasionally sewage-like smells emerge from the grey water flows in the area. To outsiders this has now become known as 'ecological smell' (Cuijpers 2006), which certainly points to the (negative) symbolic aspects the senses can have in appreciating sanitation systems (see the above discussion on the Yuck Factor).

A similar role of the senses in appreciating sanitation systems has been experienced in the case of vacuum toilets in Wageningen. Project developers who eventually had to sell the apartments with vacuum toilets to their customers had a key role in deciding whether consumers would accept the system or not. The chosen venue for one of the first project meetings with engineers, municipality and project developers, was a laboratory of the environmental engineers who were involved in the project. This lab was used for experiments with a constant flow of raw sewage water. On entering the building one of the project developers stated: "If it will smell like this, then we opt out for sure", and this scepticism has never been successfully countered.

Such negative symbolic connotations are also found in the household water case. When the first household water projects were announced the secondary water flow was commonly called 'grey water'. Not surprisingly the term invoked confusion at the side of consumers who were scared of their laundry coming out grey and although water companies quickly re-coined it into 'household water' this was only to a limited success. The term 'grey water' and its connotations of smelly, dirty water still lives on when media report on the household water projects.[6]

Lastly, concerning the sensitization of *practices* around sanitation innovations we can refer again to the composting toilets, where especially the maintenance practices were the most eye-catching. In case of the Non-Olet, after each use the waste should be covered with paper tissues, and then manually compressed with a simple device and later on disposed of in the organic waste bin. Even to satisfied users, this sensory experience was the main threshold to overcome when starting using it. How close composting toilet technology interferes with daily life practices can be illustrated with a quote from a composting toilet user in Braamwisch

[6] As late as in 2007, newspaper 'De Gelderlander' (7-12-2007 and 12-07-2007) reported on the failed household water projects in Arnhem and Wageningen by using the term 'grey water'. Interestingly, another newspaper (Utrechts Nieuwsblad 06-08-2007) uses the term 'household water' while referring to all wastewater from households.

Hamburg (Germany): "You should maintain daily. If families are not in balance, like in a divorce situation, first thing that collapses is the composting process".[7] Also in the case of vacuum toilets, and despite its rather technical and expert-driven character, the practices of consumers were high on the agenda. Especially the sensory experiences consumers would encounter, made project developers and the municipality deciding not to go on with the project. Most of their worries about consumer acceptance were expressed in terms of limited choice in toilet shapes and colours, the risk of bad smells and the noise of flushing.

3.9 Analysis of the Cases

In a first attempt to make sense about sanitation and the senses, Table 3.1 below summarizes the confrontation of our cases with the toolbox of methods and reasons behind sensitization of sanitation. Our hypotheses were that cultural display of consumption would be minimal in cases of sanitation innovation and that the method of sensitizing *practices* of sanitation would be rare. Now we can say that an equivalent to the positive display of wellness and comfort, as we have seen in sectors of energy and water supply, cannot be found in innovation projects around sanitation. Practices in sanitation do become sensitized but only in the context of a rather negative frame. Displaying consumption of sanitation services does take place, but mainly on the safe and abstract level of bigger infrastructures. At the level of flows and practices this is only done in specific, 'deep-green' niche markets for composting toilets.

Giving feedback to users – to stimulate rational resource use – by adding visibility to infrastructures, flows or practices could only be found in the household water cases

Table 3.1 Methods and reasons of sensitization in five sanitation systems in the Netherlands

Reasons	Rational resource use	Cultural display of consumption	Demonstrating sustainability to society
Methods			
Sensitizing practices	Household water	Composting toilets Non-Olet	Composting toilets Non-Olet
Sensitizing flows	Composting toilets Household water	Composting toilets Non-Olet	Composting toilets Non-Olet Household water Reed bed filters Vacuum toilets
Sensitizing infrastructures	Household water	Reed bed filters Household water	Reed bed filters Household water Vacuum toilets

[7] Site visit by author to eco village Braamwisch, Hamburg, 9 April 2003.

(double water meters and bills) and in the composting toilet case of Utrecht, where much of the data logging on flows of compost, urine and electricity was made public.

So, the majority of sanitation projects aim to demonstrate a feasible sanitation alternative to the outside world and do so by emphasising and (re)sensitizing the infrastructures as well as the flows of waste, water and waste water on the level of the public realm.

When moving from top left to bottom-right in the table the process of re-sensitizing changes from a domestic, 'private' and individual orientation into a more collective, shared, public concern. The very same technologies can contribute to processes of re-sensitization at different levels, both before and behind the real or virtual 'meter' separating the private from the public sphere.

3.10 Conclusions

There is need for a social scientific perspective on innovation in sanitation infrastructures that takes into account sensory perceptions next to the environmental and health engineering 'favourites' of closing nutrient loops, water saving, bio-energy production and pathogens removal. The notion of abstract systems that 'provide and instruct' is slowly evading in infrastructure provision and being replaced with differentiation and re-localization of production and consumption of infrastructure based services. Sanitary infrastructures are becoming 're-embedded' in the life-worlds and neighbourhoods of citizen-consumers, who are invited to actively take part in the functioning of these systems. In the context of this transition, providers and consumers have started to 'bring back the senses' into consumption and production of infrastructure-based services. We have asked ourselves whether this tendency also counts for (waste)water and sanitation service provision. Waste water systems were primarily designed in the late nineteenth century as systems to do away with the sight and smell of human waste and modern societies have grown to respond negatively towards any sight or smell related to human wastes.

The answer is that, also in sanitation innovations, the senses are indeed re-entering the scene, but in specific ways which have to be specified both for the reasons behind the sensitization process as well as regarding the levels of scale of the sanitary system. As far as sanitary systems are concerned, the sensitization process differs from what can be observed in other infrastructures of consumption. In sanitation projects, the (public) display of infrastructures and flows dominates over that of (private) practices, and (re)sensitization is mainly done to demonstrate sustainability efforts to the outside world, rather than articulating notions of cultural display or rational recourse use.

This chapter has tried to make the role of the senses in sanitation visible as a vehicle for innovation. Although bad smell doesn't sell, it is crucial not to deny or to ignore it, but instead to make a start with formulating a feasible strategy for sensitizing sanitation. Emphasizing collective aspects, working 'before the meter' and emphasizing display of sustainability primarily in relation to the public realm are elements of such a strategy. As such, the reasons behind and methods of sensitization discussed here can be regarded as a first step to make sense about 'sanitation and the senses'.

References

Beck, U. (2007). *The world at risk*. Cambridge: Polity Press.

Chappells, H., Van Vliet, B., Shove, E., Spaargaren, G., Linden, A.-L., & Klintmann, M. (2000). *Domestic consumption, utility services and the environment. Final report of the Domus project*. Wageningen: Wageningen University/Lancaster University/Lund University.

Cowan, R. S. (1983). *More work for mother: The ironies of household technology from the open hearth to the microwave*. New York: Basic Books.

Cuijpers, Y. (2006). *Verwaterend burgerschap: Technologisch burgerschap rondom een wijkwatersysteem*. Enschede: Twente University.

Estache, A. (1995). *Decentralizing infrastructure. Advantages and limitations*. Washington, DC: World Bank.

Giddens, A. (1990). *The consequences of modernity*. Cambridge: Polity Press.

Graham, S. & Marvin, S. (1995). More than ducts and wires: Post-Fordism, cities and utility networks. In P. Healy (Ed.), *Managing cities: The new urban context* (pp. 169–190). London: Wiley.

Hegger, D. (2007). *Greening sanitary systems: An end-user perspective*. Unpublished Ph.D., Wageningen University, Wageningen.

Hegger, D., Van Vliet, B., & Spaargaren, G. (2008). *Decentralized sanitation and reuse in Dutch society: Social opportunities and risks: final report for the EET-DESAR project, Wageningen, 1 January 2008*. Wageningen University, Environmental Policy Group, Wageningen.

Huizing, A. (1993). *Compost toiletten in Nederland*. Wageningen: Wetenschapswinkel Landbouwuniversiteit.

Juuti, P. S., & Katko, T.S. (Eds.) (2005). *Water, time and European cities, history matters for the future*. Tampere: EU Water Time Project.

Kessides, I. N. (2005). Infrastructure privatization and regulation: Promises and perils. *The World Bank Research Observer, 20*(1), 81.

Kristinsson, J. & Luising, A. (2001). Town planning aspects of the implementation of DESAR in new and existing townships. In P. Lens (Ed.), *Decentralised sanitation and reuse – concepts, systems and implementation*. London: IWA.

Melosi, M. (2000). *The sanitary city. Urban infrastructure in America from colonial times to the present*. Baltimore, MD: The Johns Hopkins University Press.

Post, M. (2000). *Self-sufficiency: For environmental reasons or just for fun?* Paper presented at the ESF Winter-workshop 'Infrastructures of Consumption & the Environment', Wageningen.

Russell, S. & Hampton, G. (2006). Challenges in understanding public responses and providing effective public consultation on water reuse. *Desalination, 187*(1–3), 215–227.

Shove, E. (1997). Revealing the invisible: Sociology, energy and the environment. In M. Redclift & G. Woodgate (Eds.), *The international handbook of environmental sociology* (pp. 261–273). Cheltenham: Edward Elgar.

Spaargaren, G., Mol, A. P. J., & Buttel, F. H. (Eds.). (2006). *Governing environmental flows: Global challenges to social theory*. Cambridge, MA: MIT Press.

Urry, J. (2000). *Sociology beyond society*. London: Routledge.

Van den Burg, S. W. K. (2003). Consumer-oriented monitoring and environmental reform. *Environment and Planning. C, Government & Policy, 21*(3), 371–388.

Van Vliet, B. (1995). *Waterbesparing: Over spoeling en verspilling: een vooronderzoek naar de ontwikkeling en verspreiding van diverse technologieën met als doel waterbesparing in huishoudens*. Wageningen: Wetenschapswinkel Landbouwuniversiteit.

Van Vliet, B. (2003). Differentiation and ecological modernization in water and electricity provision and consumption. *Innovation, 16*(1), 29–50.

Van Vliet, B. (2006). The sustainable transformation of sanitation. In J. P. Voss, D. Bauknecht & R. Kemp (Eds.), *Reflexive governance for sustainable development* (pp. 337–354). Cheltenham: Edward Elgar.

Van Vliet, J. (2006). *Trans(h)ition? Exploring the actor-networks constituting the arena for a transition in Dutch sanitation.* Unpublished M.Sc., Wageningen University, Wageningen.

Van Vliet, B., Chappells, H., & Shove, E. (2005). *Infrastructures of consumption: Environmental innovation in the utility industries*. London: Earthscan.

Chapter 4
Providing Sanitation for the Urban Poor in Uganda

James Okot-Okumu and Peter Oosterveer

Abstract After presenting background information on urbanization in Uganda, the chapter provides an overview of sanitation in the urban centres, where different social classes reside in separate zones. Factors determining sanitation provision and the use of sanitary facilities particularly in the informal settlements, or slums, of the larger cities are identified. Substantial groups among the population do not have access to formal sanitation facilities and have to resort to improvized unhygienic means of human excreta disposal that pose health risks. This situation underlines the need for innovative community-oriented approaches to address the sanitation challenge. By examining centralized water-based systems of sanitation vis-à-vis decentralized options, opportunities for including the urban poor in environmental service provision are identified.

4.1 Introduction

Urban sanitation constitutes an urgent environmental problem facing African urban communities and authorities. The urban population today is incomparably larger than 50 years ago, so without huge up-scaling, the centralized old urban environmental infrastructures are grossly inadequate to service all contemporary urban residents efficiently and effectively. Even maintaining and managing these existing infrastructures and services efficiently is a challenge for the local authorities in East Africa, let alone expanding them to those not yet served. However, the local urban

J. Okot-Okumu (✉)
Makerere University Institute of Environment and Natural Resources (MUIENR), P.O. Box 7298/7062, Kampala, Uganda
e-mail: jokotokumu@muienr.mak.ac.ug

P. Oosterveer
Environmental Policy Group, Wageningen University, Hollandseweg 1, 6706 KN, Wageningen, The Netherlands
e-mail: peter.oosterveer@wur.nl

B. van Vliet et al. (eds.), *Social Perspectives on the Sanitation Challenge*,
DOI 10.1007/978-90-481-3721-3_4, © Springer Science+Business Media B.V. 2010

authorities seem also unable to put in place alternative plans and regulations for physical environmental infrastructure development and their management that do cover the entire urban population. Main obstacles to substantially increase sanitation to communities in the EAC region are political instability, rapid population growth and low priority given to sanitation issues (Anschütz 1996; MoFPED 2004; WHO/UNICEF 2004). Access to sanitation in the East African urban centres ranges between 20% and 30% (UN-Habitat 2005).

Inefficient and inequitable sanitation provision is an obvious cause of the outbreaks of diseases and environmental degradation (Carr 2001). Low sanitation coverage, dilapidated waste water treatment system and prevalence of waterborne and water-related diseases were reported throughout East-Africa (East African Community 2004; LVEMP 2002; MWE 2006; Stephens and Harpham 1992; UN-Habitat 2005). The predominance of unsanitary conditions among the urban poor and neglect from urban authorities are exacerbated by the prevailing ignorance and poverty (Ababio 1992). For instance, wastes dumped by local population in storm-water drains, streams, rivers and other water points intensify environmental problems which can escalate into disastrous disease epidemics that can be a major cause for loss of productive capacity especially among the urban poor and in many instances loss of lives as well. Climate change has also been implicated in outbreaks of diseases such as cholera and diarrhoea (Hales et al. 2003).

This chapter presents an urban sanitation scenario for Uganda by identifying opportunities for the inclusion of the urban poor in waste management and sanitation. The authors claim that there are opportunities for including the poor in urban environmental service provision by developing closer connections between the conventional centralized and recently introduced decentralized systems.

This chapter begins with a general description of urbanization and sanitation in Uganda followed by an explanation of the methods used to collect data for the compilation of this chapter. The central problems in urban environmental service provision in Uganda are discussed in Section 4.3. Section 4.4 presents the policy, legal and institutional dimensions for environmental management in Uganda and innovative ways to address the problems in waste management and sanitation. After discussing aspects of funding in Section 4.5, and social and cultural issues in Section 4.6, this chapter draws some conclusions on the way forward in the final Section 4.7.

4.2 Urbanization and Sanitation in Uganda

4.2.1 Urbanization in Uganda

Uganda is a land-locked country in Eastern Africa and its current population is about 29.6 million (UBOS 2006). The country's GDP for the budgetary year 2004/5 was about 8.3 billion US$. The growth rate of its GDP was 6.2% per year signifying a per capita GDP growth of 0.2 US$ during the same budgetary year 2004/5 (UBOS 2006). The average per capita income was 280 US$ for the year 2005 (World Bank 2006).

Uganda is a tropical country, located along the equator. Except for the north-east region, the country receives an annual rainfall of more than 500 mm while the temperatures are normally around 25°C throughout the year. These warm and wet conditions in many months during the year exacerbate urban waste management and sanitation problems.

During the past two decades, Uganda experienced an increase in rural-urban migration and its urban population rate stands at 14.9% in the year 2007 (UBOS 2008). This rapid growth was not commensurate with a well-planned growth in urban services. The structure of urban communities in Uganda is complex, where the rich stay in planned housing zones but most of the poor in slums, while combined formal-informal settlements exist for the middle class. This complex structure coupled with intricate settlement patterns places serious stress on the available planning capacity and material resources, undermining the prospects for equitable and sustainable environmental service provision. Urban planning and management in Uganda is complicated by the lack of reliable data on social, economic, cultural and environment/ecological trends, leading to uninformed decisions. Local governments seem unable to effectively charge for services resulting in severely reduced financial resources (MoFPED 2004; UN-Habitat 2005) forcing local authorities to resort to external support. Financial assistance comes mainly from the central government, Non-Governmental Organizations (NGOs), Community Based Organizations (CBOs), and international donors (MoLG 2007). Such external financial support is in most cases conditional and directed towards a particular sector, whereby funding for sanitation is often not prioritized (Section 4.3) compared for example to water provision. Sanitation services are unbalanced with regard to the social classes and the spatial segments within the cities served. Hence, the urban poor living in the (peri-)urban slums do not receive adequate sanitation services from local governments while they are constrained by their limited capacity to secure these services themselves.

To increase accessibility of the poor and strengthen sustainability, the introduction in urban development strategies of a Modernized Mixtures MM (Spaargaren et al. 2006) approach should be welcomed. The MM approach (Oosterveer and Sano 2008) stands for an integrated approach, taking the best from both existing sanitary strategies (centralized and decentralized) in order to better fit the particular local situations in both social and technological respects (see Chapter 2). Decentralized sanitation and Resource Reuse (DeSaR) systems look promising especially for the poor communities that can subsidise their expenditures by growing their own food (urban agriculture) and closing the water and matter loops. In exploring the potential of alternatives, not only the physical infrastructures should be covered but the relevant political, social and institutional arrangements, as well.

In this chapter the results from a study on sanitation systems in urban centres of the four administrative regions[1] of Uganda are presented. These urban centres include Kampala City, the municipalities of Lira, Arua, Soroti, Mbale, Masaka, Kabale and Fort Portal, and Mayuge Town Council (Fig. 4.1).

Urban centres in Uganda include cities, municipalities and town councils, whereby the grading is based on the number of inhabitants. A town council must

[1]Central, Eastern, Northern and Western.

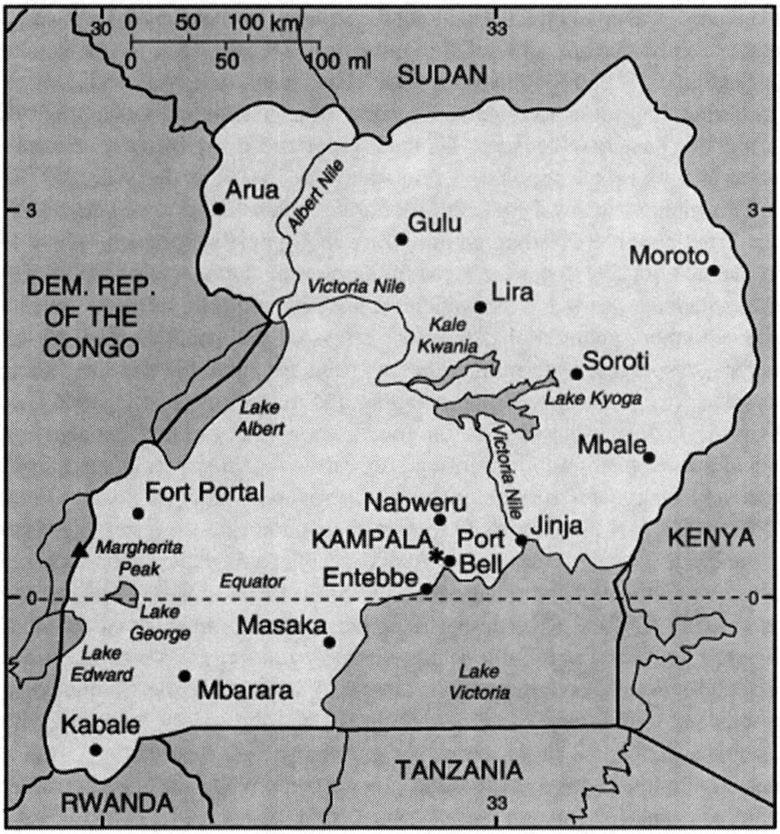

Fig. 4.1 Map of Uganda showing the major urban areas (http://www.idrc.ca/openebooks/189-2/f0032-01.gif)

have a population of more than 25,000 inhabitants, a municipal council more than 100,000 inhabitants and a city council a population of more than 500,000 inhabitants (Local Government Act 1997, Schedule 3, Article 32–1).

4.2.2 *Study Methods*

The findings reported in this chapter are based on field research and a review of existing literature aimed at identifying and assessing the sanitation provision mechanisms and the problems related to their implementation in the urban centres of Uganda. Both quantitative and qualitative data collection methods[2] were used in the field study. The methods include community surveys through the administration of questionnaires to randomly sampled households to acquire information about the

[2]Research methods are based on Aagaard-Hansen and Yoder (2007).

sanitation condition and on the relevant social and economic factors (yielding both qualitative and quantitative data). In addition, focus group discussions (FGDs) were performed with local government authorities and staff, and with NGOs on problems related to sanitation and waste management in urban areas (yielding qualitative data), observations were made on the existing sanitation facilities determining type, location, operational conditions, environmental impacts and opportunities for improvement (qualitative data). These findings were supplemented with document reviews (generating both qualitative and quantitative data). The lead researcher together with trained enumerators administered the questionnaires and conducted FGDs from January to April 2008.

4.3 Central Problems in Urban Environmental Service Provision

The provision of sanitary services and the related challenges are discussed here while distinguishing different social groups found in the urban centres of Uganda. The principal two distinct groups are the (peri-) urban slum dwelling poor staying in unplanned high-density communities and the affluent class staying in well-planned residential areas. These (peri-) urban communities consist of a mixture of almost all the tribes of Uganda and this cultural mix results in a complex social structure (Okot-Okumu 2008). Among the low-income poor households in these slums there is lack of water and sanitation (15–45 persons per latrine) and solid wastes are in most cases left to rot uncollected. Contrary to water projects, solid waste and sanitation facilities in these neighbourhoods are less utilized and less well-maintained except when the communities accept them based on culture, cost and technology (see Section 4.6). This is because water is a basic necessity of life and as communities are immediately strained without it, this is therefore their first choice. A poor community even if aware of the linkages between good sanitation, environmental pollution and human health will show lower priority to waste management and sanitation compared to other basic necessities. However, many inhabitants of poor neighbourhoods do not clearly perceive the connection between latrine usage and health (Cairncross and Feachem 1993). The desire for privacy, convenience or social status is usually more effective in generating demand than health concerns. At the same time sanitation laws in Uganda are deficient and the implementation is weak.

Although in this study local government staff rated solid waste management and sanitation first and second respectively as the major urban environmental problems, urban authorities give these issues low priority in their planning and resource allocation (MoLG 2007; Okot-Okumu 2008). This is because the technical staff does not determine or significantly influence resource allocation. Rather resource allocation is influenced by the executive staff that in many cases lacks appreciation of the importance of community sanitation and its relevance for welfare and productivity.

Effective mainstreaming environmental management in the urban policies and plans is still lacking (MoLG 2007; Okot-Okumu 2008). Therefore, since 2005, there is ongoing training in environment and natural resources management organized by the Ministry of Local Government (MoLG) for all local governments in Uganda. An important aspect of this training is the mainstreaming of environment, gender and equity issues in policies and plans for districts and urban centres (Okot-Okumu 2008). In addition, the MoLG produced in the year 2007 environmental management information sheets and posters for the local governments. The training and other awareness-raising materials are developed for skill development of the LG's technical staff while information sheets and posters are specifically designed to target the community. The environmental management training and information materials however all do lack the key issue of partnership which is vital for community involvement.

Currently the official coverage of improved sanitation for the Ugandan population is 92% in urban and 62% in rural areas (MWE 2007; NEMA 2005). This estimated coverage is based on access to latrines and hand-washing at the household level, which overestimates effective access to safe sanitation. More elaborate studies, such as Mubiru (2000) and MoWLE (2004), indicate safe coverage at an average of only about 30%. Many of the latrines are shared by large numbers of people (MWE 2007) whereby management responsibilities are not very clear, making them dirty and structurally poor in terms of health and safety (Table 4.1).

Table 4.1 Conditions that make sanitation facilities unsafe[a]

Condition of facility and method of data collection	Health and safety implication
Unclean with faeces and wet floor, bad smell, no pit hole cover (observation – qualitative)	Risk of disease transmission by direct contact and by transmission agents like flies
Flies within and around the sanitation facility, cockroaches and rodents	Risk of disease transmission by the observed agents
No sanplats or slabs	Risk of accidental collapse
Too close to housing and water points (observation, questionnaires and FGD-qualitative)	Risk of water pollution, disease transmission
Dilapidated state of the sanitary structures(observation – qualitative)	Risk of environmental pollution, lack of safety for users, disease transmission
Sullage from bathing and urinal shelters discharged in open drains, grey water pools(observation, questionnaires and FGD-qualitative)	Risk of disease transmission by contact contamination especially among children, environmental contamination of water points
Faecal matter disposed in polyethylene bags, faeces littering compounds and waste dumps(observation, questionnaires and FGD- qualitative)	Risk of disease transmission by contact contamination especially among children, stray animals and environmental contamination including water points
No hand washing facilities at latrines (observation, questionnaires and FGD- qualitative)	Risk of human-to-human contamination

[a]Based on observations from the field study collected between January and April 2008. See Section 4.2.2 for the methods used

The sewage network in towns is very small and covers only 7% of the national urban population (see Letema et al., Chapter 9 in this volume). These networks are restricted to the central business districts of the larger cities and to affluent residential areas close to the centre of these towns. The sewer networks have never been expanded to the newly built residential areas where septic tank-soak pits are necessarily used by the richer populations. The septic tanks are emptied by cesspool emptying trucks that transport the waste to sewage treatment points. Large housing estates in (peri-) urban areas use sewage lagoons or maturation ponds that drain into the environment, usually adjacent wetlands.

No attempts have ever been made to connect the (peri-) urban slum areas where most of the poor live because of fear that cost recovery will fail as the poor residents lack the means to afford these services. There are however many smaller towns including the secondary towns close to Lake Victoria that do not have sewer lines at all (UN-Habitat 2005). On-site sanitation systems that include septic tank-soak pit systems, traditional pit latrines, improved pit latrines (e.g. VIP, slabs, sanplats, etc), and some flush toilets account for 90% of the sanitary facilities in these urban centres (Fig. 4.2). However, on average only about 30% of these facilities are managed hygienically. LVEMP (2002) noted that very poor sanitation and solid waste management in secondary towns in Ugandan results in a 20% effective latrine coverage only. Most human waste in these locations is discharged directly into the environment.

The existing urban sanitation infrastructures are in most cases in a state of disrepair. Traditionally, managing the sanitary infrastructures and services was the responsibility of the municipal authorities, in particular the Department of Public Health. The recently created Department of Environment coordinates and supervises environmental management at local government level, but has little practical influence on waste management and sanitation because the Department lacks the necessary resources (e.g. staff, funds, equipment). Waste and sanitation was generally not considered a high priority and revenues were regularly used to fund other 'more urgent' expenditures (UN-Habitat 2005). Therefore little or no money is spent on the maintenance or invested in the expansion and renewal of infrastructures.

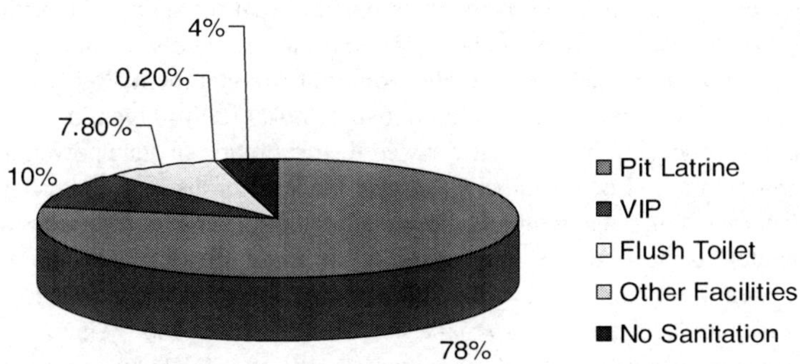

Fig. 4.2 Percentage distribution of types of sanitation facilities in Ugandan urban centres (Okot-Okumu 2008)

In combination with the growing presence of silt, the lack of proper maintenance, misuse of toilet facilities and indiscriminate disposal of solid wastes are causing degradation of existing infrastructures and local flooding and environmental pollution. This is the scenario in most urban areas of Uganda.

A reform study for urban water and sanitation was done between September 1999 and December 2000. The resulting reform in water and sanitation was a clear effort to offset the negative institutional weaknesses whereby services have now been commercialized and assets are held by separate (parastatals) entities while all stakeholders are represented to ensure an integrated approach in management. In Uganda this approach is spearheaded by the National Water and Sewerage Corporation (NWSC) in 18 towns. NWSC is mainly concerned with sewer line provision, which is not affordable to the urban poor.

The integrated approach is however being practiced well only at the central government level. The strength of this integrated approach dwindles as one moves down the ladder towards the lower government levels where the poor communities can be found. Community involvement in waste management is discussed in depth by Muller and Hoffman (2001) and can be adopted for sanitation programmes as well. To be successful, however, all aspects of management – technical, environmental, financial/economic, socio-cultural, institutional and policy (including political) – should be included.

4.4 Policy, Legal and Institutional Framework for Sanitation

4.4.1 Policy and Legislation

In Uganda, environmental policymaking remains the function of the central government, but implementation of its policies and regulations is passed on to the districts. This is in line with the general decentralization process that has been adopted by the country.

The constitution of Uganda (Government of Uganda 1995; clause 39) states that *'every Ugandan has a right to a clean and healthy environment'* while at the same time every citizen is expected to play his part in creating such a situation (clause 17[j]) – *'it is the duty of every citizen of Uganda to create and protect a clean and healthy environment'*. The National Environment Management Policy (NEMP) of 1994 Section 3.9 sets the goal for the control of pollution in Uganda.

The policy of decentralization has allowed the devolution of state powers from central government to the Local Councils (LCs) at the level of districts and urban authorities. Environmental management and planning, including sanitation services and waste management, has become the responsibility of these districts and municipalities. Therefore local governments have the full mandate for solid waste management and sanitation and the legal authority to make specific ordinances and by-laws.

The National Environment Management Authority (NEMA) is empowered to establish environmental standards and regulations (e.g. sanitation and solid waste management regulations) and to coordinate and supervise environmental management in Uganda. The Public Health Act Cap 281 of 2000 requires local authorities to take measures for maintaining their area in a clean and sanitary condition at all times to prevent the outbreak of diseases and to safeguard and promote public health. While these regulations exist, so far no urban authority has implemented them effectively. Treatment of sewage remains inadequate or is completely absent while solid waste is disposed of indiscriminately, polluting the environment and causing health risks. There is lack of capacity for implementing sanitation policies and regulations. The presence of several policies relevant for the sanitation sector also causes implementation problems through duplication of activities within other sectors.

Some districts and lower level local governments lack up-to-date ordinances and by-laws to handle issues in sanitation and solid waste management. Community survey and focus group discussions found low levels of use of the regulatory documents by the local governments. It was stated that these documents are difficult to read or understand. In such cases an abridged version of these documents would be more helpful.

Political interference is a major problem in African countries impacting on environmental management as reported by Palczynski (2002) and Rotich et al. (2006). Politicians may interfere with environmental management projects by making statements that are contradictory to the official environmental management policy and encourage communities to refuse cooperation and contribution in sanitation projects. This interference weakens environmental management institutions' ability to implement laws and regulations and makes the local community difficult to deal with.

When reviewing the existing policy and regulations we observe that despite the presence of many policy and legal or regulatory documents, urban centres in Uganda still experience very poor sanitation and waste management. The making and implementation process of these statutory instruments is failing particularly because not all relevant stakeholders are involved. These failures are most felt by the poor especially in the congested slums. It is therefore important to involve the most impacted community because every neighbourhood may have its unique and complex mixture (e.g. social, economic, cultural, and religious) that requires attention for policy and environmental regulations. The opportunity for the poor to participate in the making of policies and regulations is recognized in the existing Acts of parliament mentioned above but not fully implemented. To overcome this limitation the poor urban community should be involved in the revision of these statutory documents by engaging them together with local policy/legal experts in analysing their neighbourhood waste and sanitation problems and drafting their own vision on waste management and sanitation. Community views on improving policies and regulations on waste and sanitation should be submitted for consideration by the next higher level of local government. Abridged versions of the existing statutory documents should be produced for easy interpretation and use by all stakeholders. An example of this is the version of the National Environment Statute/Environment Act of 2005 termed *'Popular Version'* that is more receptive for use by stakeholders.

NGOs should support the poor in making their own by-laws on wastes and sanitation. There is already a strong contribution from NGOs through partnerships in urban waste management and sanitation in Uganda (Tukahirwa et al. 2008). The policy and regulation making process should be opened up from the very beginning to enable effective community contribution and acceptance during implementation and to develop community responsive policies and laws on waste and sanitation.

4.4.2 Institutional Arrangements

The Ugandan National Environment Act (Cap 153) and the Local Government Act (Cap 243) specifically decentralizes environmental management to local governments to increase its cost-effectiveness and community involvement. Urban councils/authorities are responsible for decentralized services that include sanitation and waste management. The institutional framework for water and sanitation management in Uganda is displayed in Fig. 4.3.

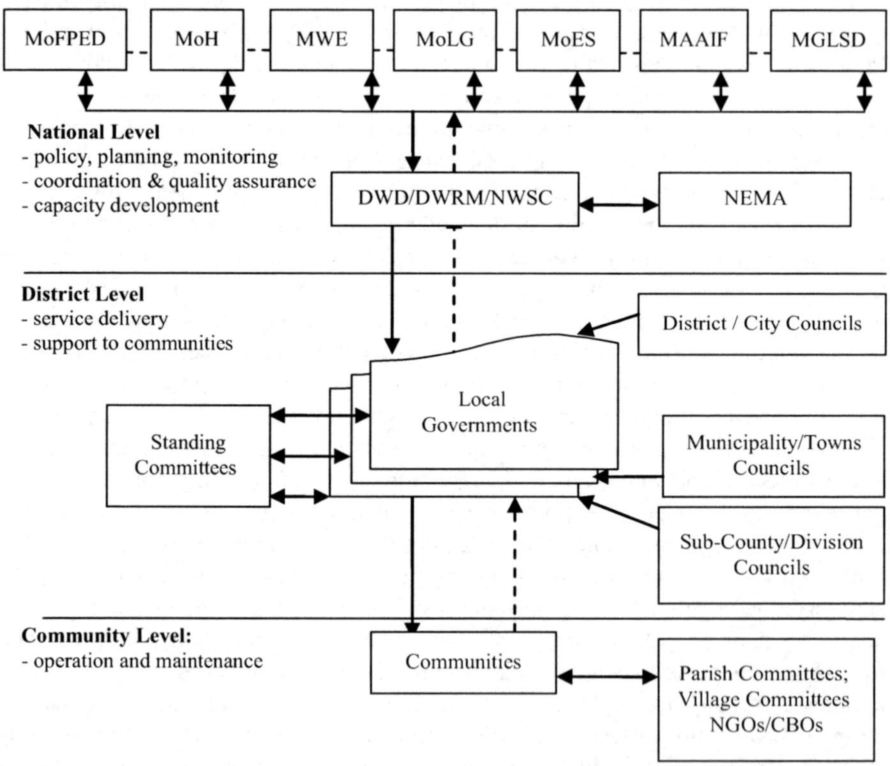

Fig. 4.3 The institutional framework for water and sanitation management. MoFPED: Ministry of Finance Planning and Economic Development, MoH: Ministry of Health, MWE: Ministry of Water and Energy, MoLG: Ministry of Local Government, MoES: Ministry of Education and Sports, MAAIF: Ministry of Agriculture Animal Industry and Fisheries, MGLSD: Ministry of Gender Labour and Social Development, NEMA: National Environment Management Authority, DWD: Directorate of Water Development, DWRM: Directorate of Water Resources Management Environment, NWSC: National Water and Sewerage Corporation

At the national level there is the Water and Sanitation sector Working Group (WSSWG) that is comprized of the different relevant Ministries, the international development partners, NGOs and local governments. WSSWG has two sub-sector working groups and one of them is on sanitation. The role of WSSWG includes providing policy and technical guidance to the sanitation sector, preparation of medium term budget framework papers for the sector, preparation of annual sector performance reports and ensuring the implementation of Sector Wide Approach (SWAp) in the sanitation sector.

At the local level, also District Water and Sanitation Coordination Committees have been established to strengthen collaboration and coordination between the different policy areas (health, education, agriculture and social development), the private sector, NGOs, CBOs and civil society. NGOs and CBOs are very active in the provision of water and sanitation services, in particular through the construction of facilities, capacity building of local governments and community mobilization, hygiene promotion, advocacy and lobbying. Currently some 200 NGOs and CBOs are involved in water and sanitation activities in Uganda (MWE 2007).

Regional and international organizations that could contribute to further collaboration in the domain of water and sanitation in the East African region are UN-Habitat, the Lake Victoria Basin Commission (LVBC), the Nile Basin Initiative (NBI), the Lake Victoria Environment Management Project (LVEMP), the Lake Victoria Local Authorities Cooperation (LVRLAC) and the East African Communities Organization for the management of Lake Victoria (ECOVIC). Despite the presence of this large number of organizations, there is little effective cooperation and harmonization of water and sanitation efforts, although this could go a long way in addressing the many challenges.

4.4.3 Institutional Innovations in the Water and Sanitation Sector

In Uganda the management of urban water supply and sanitary infrastructures are closely linked and the planning is guided by one integrated policy document. In order to meet the targets of the Poverty Eradication Action Plan (PEAP), MoFPED (2004) initiated a reform of the water and sanitation sector. This is in line with the MDG No 7, targeting improved sanitation for the poor. The sector reform process in Uganda was developed through a participatory approach with strong links to the objectives of poverty alleviation. The water and sanitation reform process involved government lead ministries, civil society, NGOs and international donor agencies (e.g. DANIDA, SIDA). The Uganda Water and Sanitation NGO Network (UWASNET) is an umbrella organization and has played a great role in coordinating, strengthening information sharing and networking among the NGOs and CBOs operating in this domain and with the government. The aims of the sector reform were to ensure that water supply and sanitation services are provided with increased performance and cost-effectiveness, and to reduce the government's financial burden without compromising the provision of equitable and sustainable services.

To implement these objectives asset holding and development were separated from service provision and regulatory activities, while decentralization and devolution of responsibilities, improved transparency, reduced political interference and effective public-private partnerships would all contribute further.

The first step in the reform process was the enactment of the 1999 National Water Policy (NWP) to strengthen the regulatory framework and provide a basis for cost recovery, including the introduction of commercialized operations. Based on a sub-sector study in the year 2000, public-private partnership was recommended whereby the ownership of assets involved in urban water supply and sanitation is retained by the public sector and service delivery is done by private operators. For small towns both responsibilities remain with the local governments. While the larger towns (18) are guided by the NWSC, the smaller towns (48) are guided by the (national) Directorate of Water Development (DWD).

The reform process also generated a consensus on the importance of a sector-wide approach (SWAp), comparable to the strategy that was already adopted with some success by the Health and the Education sectors. Such a sector-wide approach aims at moving towards comprehensive programs that are well coordinated in funding and eliminate duplication of efforts. There is recognition by the government that as the reforms are being implemented there is an increasing need to co-ordinate its own efforts with those of its development partners.

We may observe from these findings that although the reforms generate positive results there are weaknesses in the decentralization and devolution processes because local governments often lack the capacity required to successfully take their responsibilities. Local governments still depend largely on the central government and external support (MWE 2007). This problem has resulted in the community not being well represented within the institutional structure even though this is provided for by law as displayed in Fig. 4.3. The existing institutional structures should be modified after consulting the community so that the poor are adequately and effectively involved in the planning and implementation of waste and sanitation programmes in their locality. Institutionalization of the roles of the community in the long-term plans for waste and sanitation should be done to enable their integration in the overall strategy, which may be expressed in the form of laws, regulations or relevant documents.

A key area for involving the community and the poor is the introduction of decentralized systems. DeSaR systems aimed at closing the water and matter loops and at turning waste into goods for the poor communities can be implemented because this is already supported by the decentralization policy. However, this model can only be successful if the community sees tangible benefits from the waste itself or from the waste management and sanitation activities. A study in Kampala by Niwagaba et al. (2008) used social marketing techniques to promote sanitation and was successful in developing awareness and skills among the community.

The community can also be involved through cooperatives or businesses (e.g. micro- and small-scale enterprises) as is ably discussed by Scheinberg (2001), which, apart from being an effective means of solving waste and sanitation problems, is also a job creator. Small-Scale Enterprises can take tasks such as waste collection, emptying latrines and septic tanks, composting and recycling. Recognition of these enterprises by the LGs is important and initial support by agencies like NGOs may be vital for these entrepreneurs to succeed.

4.5 Funding

The financial resources available to the urban authorities for managing water and sanitation originate from the central government, international donor agencies and the locally mobilized revenues. Section 81 of Local Governments Act 1997 provides that local governments may levy charges and collect fees and taxes, while the rates of these levies are regulated through the Local Government Rating Act (MoLG 2003). Suspension by the national government in 2005 of the Graduated Tax which was a major source of revenue for local governments has complicated their funding possibilities and they have not been able to raise adequate revenues. Moreover, the sector of sanitation and waste management receives even less funding from the local governments because this is considered a low priority. In the Financial Year 2006/07, less than 10% of the total municipal budget was used for waste management and sanitation by local governments in Uganda (MWE 2007). Most of the funding is therefore coming from international donors, NGOs and conditional government grants through the Local Government Development Programme. This is not a sustainable system of financing because it has no local resource base, while external funding sources (e.g. donors, NGOs, LGs) tend to have prescribed priorities which are not necessarily those of the poor. For sustainability significant funds must accrue from the community initiated activities.

It can be concluded that there is an opportunity for partnership in waste management and sanitation between the local authorities, NGOs and the community. The community can form neighbourhood waste and sanitation management associations that are initially supported financially by LGs or NGOs to provide a revolving fund. Some of the income-generating opportunities in the waste and sanitation sectors are solid waste sorting, collection, composting, reuse, recycling, briquette making, handicraft making from wastes and emptying of pit latrines and septic tanks. These activities are already taking place in Uganda especially in the larger urban areas like Kampala, Jinja and Entebbe. However, many projects collapse because of their failure to recover costs. Professional expertise within the LGs and NGOs could be used to boost the community potentials (technical, financial) and promote sustainability. Apart from improving on waste and sanitation management for environmental health this will also cut the funds required from external sources. Ownership of waste and sanitation projects by the community can provide the opportunity to involve the poor more effectively in waste management within their own neighbourhood. The tangible rewards can attract more poor individuals and the idea can also be easily duplicated in other neighbourhoods.

4.6 Social and Cultural Issues

Poor sanitation is exacerbated by several social and cultural factors. These factors include poor knowledge and awareness of the health and environmental implications of poor sanitary practices while particular cultural beliefs and practices can hinder the introduction of innovations in sanitation technologies, systems and practices. A general lack of community participation complicates the adequate inclusion of and response to these issues, slowing down the modernization process (Ajello 2006; Mubiru 2000). These social and cultural issues, in combination with the absence of sanitation facilities for the poor and the presence of congested housing in many cities, result in the adoption of bad practices like uncontrolled dumping of solid waste and faeces. This has probably caused the recent cholera and diarrheal disease outbreaks in slums and camps for internally displaced people (MoH 2007, 2008). There were cholera and diarrhoeal disease outbreaks in Pabbo camp in the north of Uganda in the year 2005, Bundibugyo, Hoima and Kibaale in western Uganda in the year 2006, and in Kampala's (peri-) urban areas in the years 1997/98, 2006, 2007. These outbreaks were also linked to the onset of the rainy season and to floods.

The field study of Ugandan towns by the authors found that in some neighbourhoods cultural beliefs prevent the (effective) use of the available sanitary facilities, in particular among the poor in (peri-) urban areas. Some ethnic groups have taboos that prevent realizing the potential benefits from improved sanitation infrastructures and services. For example, some groups in Uganda do not allow in-laws to share latrines and even forbid pregnant women to use them at all (Mubiru 2000; see also Ajello 2006; Tanner 1995). In some cases, these negative attitudes and cultural taboos prevent particular groups and people to actively participate in public campaigns and other activities to improve sanitary infrastructure (Palczynski 2002). Higher levels of education and social status were found to have positive impacts on sanitation behaviour (Ajello 2006; Mpamize 1998; Mubiru 2000; Omer 2004). Innovative methods that could be used to involve the community better in sanitation management and ensure sustainability are summarized in Table 4.2. The table was constructed from data collected through questionnaires administered to the community, interview of key informants and focus group discussions. This table makes perfectly clear that concrete perspectives exist for engaging the local community in improving the sanitation sector in Uganda.

4.7 Conclusions

The rapid urbanization and population growth in Uganda together with increased influx from the rural areas is creating mushrooming (peri-) urban informal settlements. It is difficult to provide environmental services for such poor communities because of the weak or absent infrastructures. Traditional approaches to waste management and sanitation have their limitations, and consequently the suitability

Table 4.2 Possible community-oriented innovative methods for sanitation programmes

Innovative methods for sanitation improvement	Expected impact
Formation of community health and sanitation committees	Sense of ownership of the projects Increased acceptance and participation in projects Increase involvement of the poor in the planning and implementation of waste and sanitation programmes Increased potential for functional sustainability of waste and sanitation facilities
Introduction of community entrepreneurship in waste and sanitation (e.g. construction of facilities, maintenance, collection and disposal, composting, briquette making, handicrafts, recycling)	Alternative source of livelihood (in line with poverty reduction strategy) More effective waste and sanitation systems (cleaner, safer facilities), cost recovery Better waste management and sanitation, cleaner and safer environment to live in
Community targeted programs (e.g. school sanitation programs, hand washing campaigns, neighbourhood health and hygiene programs)	Gradual positive change in community hygiene and sanitation, attitude and behaviour by targeting children and the youth Increased sanitation coverage and sustainable usage of facilities Positive health impact
Designing sanitation facilities involving the community whereby the current centralized and decentralized systems complement each other	A more flexible system that can be adapted for a particular community, geography, climate Increased use-sustainability Cost effective and affordable by the community
Capacity building of local governments (e.g. training, awareness programs, production of learning materials [e.g. information sheets, posters])	Improved leadership, coordination and supervision capabilities More understanding of needs for sanitation Mainstreaming of sanitation in development plans

of the already implemented infrastructure to respond to current needs decreases (Scheelbeek 2006). In most cases traditional approaches do not respond to requirements expressed by the community or neighbourhood, resulting in rejection and improper use of the facilities and therefore to non-sustainability. Adopting the large-scale European or American centralized sewerage systems in African circumstances is often unsuccessful (LVEMP 2005; MoLG 2007; MWE 2007; Scheelbeek 2006; UN-Habitat 2005). These systems require financial means for construction and maintenance as well as planning and management capacities that go beyond the available resources. On the other hand, decentralized systems such as single household pit latrines or septic tanks that are 'scattered' around the neighbourhood cannot provide the answer either. These facilities saturate a neighbourhood with sewage collection points, thus increasing the risk of pathogen transmission compared with centralized systems that exhibit less human contact with waste flows. Decentralized systems also require the application of strict criteria in site selection and management to avoid seepage of contaminants into ground and surface water, and insect and rodent infection of home facilities. The resulting appalling waste situations in neighbourhoods cause pollution of water sources (Howard et al. 2003;

Nasinyama et al. 2000; Nsubuga et al. 2004; Skoog 2004; Taylor and Howard 1995) when rain carries pollutants from wastes into water sources (Howard et al. 2003; Taylor and Howard 1995).

The official commitment of the national authorities offers a starting point for initiating community-based strategies and implementation practices. This official engagement should be expressed in creating an adequate institutional environment for pro-poor sanitation programs and create opportunities for investment in sanitation infrastructures and services. The interest in this domain from private entrepreneurs, NGOs, CBOs and international donor agencies should be capitalized upon. Ultimately this engagement will result in more diverse and flexible opportunities for sanitation, that better fit the conditions of the poor, in particular their limited ability to contribute, their illiteracy and absence of (a culture of) sanitation. Local authorities should be pressurized through local communities and national authorities to give priority to this policy domain and secure the necessary financial and administrative resources to address a problem that is already urgent and will only become more so in the future as a result of rapid population influx.

Involving the urban poor in improving waste management and sanitation should be approached in an integrated manner that considers other stakeholders (NGOs, CBOs, LGs, entrepreneurs) as partners and includes all the aspects (environment, technical, financial socio-cultural, institutional, policy) of waste management and sanitation. This chapter identified key entry points for the urban poor in waste management and sanitation, showing the possibilities in the policy (legal, political), institutional and funding domains to improve the level of performance and success. This illustrated the existence of opportunities for the involvement of urban poor in waste management and sanitation that are not yet adequately exploited. Already introduced local solutions have to be considered as part of larger systems that best fit the urban neighbourhoods and the urban area as a whole.

Targeting the Millennium Development Goal 7 and PEAP covering water and sanitation demands an integrated approach with strong community participation. Innovative concepts and strategies, such as the Modernized Mixture approach (Spaargaren et al. 2006), offer better perspectives to broadly consider the community, engage all stakeholders, acknowledge the specific local conditions and incorporate ecological sanitation. It is important for an African country, such as Uganda, to not only aim at providing sanitation services that deal with the problem of excreta disposal, but to also provide services that are sustainable in the long term as observed by Carter and Rwamwanja (2006).

References

Aagaard-Hansen, J., & Yoder, P. S. (2007). Buying research, customers guide. DBL-Centre for Health Research and Development. Copenhagen: University of Copenhagen.

Ababio, B. M. (1992). *Urban primary health care: Accra healthy cities project (draft)*. Accra, Ghana: Accra Planning and Development Program.

Ajello, G. O. (2006). *The challenge of implementing ecological sanitation in Kampala. The community perspective.* Kampala: Makerere University.

Anschütz, J. (1996). *Community-based solid waste management and water supply projects: Problems and solutions compared. A survey of literature.* Gouda: WASTE.

Cairncross, S. & Feachem, R. G. (1993). *Environmental health engineering in the tropics: An introductory text.* New York: Wiley.

Carr, R. (2001). Excreta related infections and the role of sanitation in the control of transmission. In L. Fewtrell, & J. Batram (Eds.), *Water quality: Guidelines, standards and health. Assessment of risk and risk management for water-related infectious disease* (pp. 89–114). London: IWA Publishing.

Carter, R. C., & Rwamwanja, R. (2006). *Functional sustainability in community water and sanitation. A case study from South-west Uganda.* Kampala: Church of Uganda/Diocese of Kigezi/ Cranfield University/Tearfund.

EAC (East African Community). (2004). *Basic information on Lake Victoria issues.* Retrieved March 30, 2009, from http://www.eac.int/lvdp. Accessed 30 March 2009.

Government of Uganda. (1995). *Constitution of the republic of Uganda.* Kampala: Government of Uganda.

Hales, S., Edwards, S. J., & Kovats, R. S. (2003). Impacts on health of climate extremes. In A. J. McMichael, D. H. Campbell-Lendrum, C. F. Corvalan, K. L. Ebi, A. K. Githeko, J. D. Scheraga, et al. (Eds.), *Climate change and human health: Risks and responses* (pp. 79–102). Geneva: World Health Organisation.

Howard, G., Pedley, S., Barrett, M., Nalubega, M., & Johal, K. (2003). Risk factors contributing to microbiological contamination of shallow groundwater in Kampala, Uganda. *Water Research, 37*, 3421–3429.

LVEMP. (2005). *Water quality synthesis report.* Entebbe: LVEMP.

LVEMP (Lake Victoria Environmental Management Project). (2002). *Disease prevalence and sanitary conditions at landing sites of Lake Victoria.* Entebbe: LVEMP.

MoFPED (Ministry of Finance Planning and Economic Development). (2004). *Poverty Eradication Action Plan.* Kampala: Government of Uganda.

MoH. (2008). *Annual health sector performance report.* Kampala: Government of Uganda.

MoH (Ministry of Health). (2007). *Annual health sector performance report.* Kampala: Government of Uganda.

MoLG. (2003). *Local Government rating Act.* Kampala: Government of Uganda.

MoLG (Ministry of Local Government). (2007). *Environment and social management framework. Local Government Management and Service Delivery Program (LGMSDP), final report.* Kampala: Government of Uganda.

MoWLE (Ministry of Water Lands and Environment). (2004). *Sanitation strategy and master plan for Kampala city.* Kampala: Government of Uganda.

Mpamize, S. (1998). *Domestic solid waste management in Nakawa East division, Kampala.* Kampala: Makerere University.

Mubiru, W. (2000). *Human excreta disposal and hygiene practices in Mbarara district, Uganda.* Kampala: Makerere University.

Muller, M. & Hoffman, L. (2001). *Community partnerships in integrated sustainable waste management. Tools for decision-makers.* Gouda: WASTE.

MWE. (2007). *Water and sanitation sector performance report.* Kampala: MWE.

MWE (Ministry of Water and Environment). (2006). *The year book of water resources management department 2004–2005.* Kampala: Ministry of Water and Environment.

Nasinyama, G. W., McEwen, S. A., Wilson, J., Walter-Toews, D., Gyles, C. L., & Opuda-Asibo, J. (2000). Risk factors for acute diarrhoea among inhabitants of Kampala District, Uganda. *South African Medical Journal, 90*, 891–898.

NEMA (National Environment Management Authority). (2005). *State of environment report for Uganda (2004/2005).* Kampala: Government of Uganda.

Niwagaba, C. B., Ssemanda, J., Sande, H. K., & Kamara, I. (2008). *Social marketing for sanitation improvement in Kampala, Uganda.* Wageningen: IWA International Conference, Sanitation Challenges.

Nsubuga, F. B., Kansiime, F., & Okot-Okumu, J. (2004). Pollution of protected springs in relation to high and low density settlements in Kampala-Uganda. *Physics and Chemistry of the Earth, 29*, 1153–1159.

Okot-Okumu, J. (2008). *Issues and challenges of sanitation provision in urban centres of uganda*. Wageningen: IWA International Conference, Sanitation Challenges.

Omer, M. M. Y. (2004). *Community willingness to pay for solid waste collection and disposal in Nakawa division. Kampala. District*. Kampala: Makerere University.

Oosterveer, P. & Sano, J. (2008). *Transition towards developing sustainable urban environmental infrastructure in East Africa*. Wageningen: IWA International Conference, Sanitation Challenges.

Palczynski, J. R. (2002). *Study on solid waste management options for Africa*. Abidjan: African Development Bank.

Rotich, H. K., Yongsheng, Z., & Jun, D. (2006). Municipal solid waste management challenges in developing countries: Kenyan case study. *Waste Management, 26*, 92–100.

Scheelbeek, P. (2006). *Urban environmental infrastructure around Lake Victoria: Challenges and opportunities of decentralized sanitation systems for the urban poor*. Wageningen: Wageningen University.

Scheinberg, A. (2001). *Micro-and small enterprises in integrated sustainable waste management*. Gouda: WASTE.

Skoog, K. (2004). *Waste management in Kampala, Uganda and the impact of Mpererwe Landfill. (Minor field study No.274)*. Uppsala: Swedish University of Agricultural Science.

Spaargaren, G., Oosterveer, P., Van Buuren, J., & Mol, A. P. J. (2006). *Mixed modernities: Towards viable urban environmental infrastructure development in East Africa*. Wageningen: Wageningen University.

Stephens, C. & Harpham, T. (1992). Health and environment in urban areas in developing countries. *Third World Planning Review, 14*, 267–282.

Tanner, R. (1995). Exerting, execrators and social policy. Cross-cultural observations on under-researched activities. *Journal on Preventive Medicine and Hygiene, 36*, 1–10.

Taylor, R. G., & Howard, K. W. F. (1995). Averting shallow-well contamination in Uganda. In: *Sustainability of water and sanitation systems; 21st WEDC conference* (pp. 62–65). Kampala.

Tukahirwa, J. T., Mol, A. P. J., & Oosterveer, P. (2008). *Participation of NGOs and CBOs in urban sanitation and solid waste management in Uganda*. Wageningen: IWA International Conference, Sanitation Challenges.

UBOS (Uganda National Bureau of Statistics). (2006). *Uganda National Household Survey 2005/6. Report on socio-economic module*. Kampala: UBOS.

UBOS. (2008). *Statistical abstract*. Kampala: UBOS.

UN-Habitat. (2005). *Programme proposal (revised version, January 2005) for the Lake Victoria region water and sanitation initiative. Supporting secondary urban centres in the Lake Victoria region to achieve the millennium development goals*. Nairobi: UN-Habitat.

WHO/UNICEF, Joint Monitoring Programme for Water Supply and Sanitation. (2004). *Meeting the MDG drinking water and sanitation target: A mid-term assessment of progress*. Geneva: WHO/UNICEF.

World Bank. (2006). Uganda data profile. In: *World development indicators database*. Washington, DC: World Bank.

Part II
Decision-Making Tools

Chapter 5
A Flowstream Approach for Sustainable Sanitation Systems

Elizabeth Tilley, Christian Zurbrügg and Christoph Lüthi

Abstract Reaching the Millennium Development Goals for Sanitation is a challenge. To address this challenge, numerous technological innovations have been developed. But with so many innovations and a wide range of existing technologies appropriate in different settings, difficulties with communication and knowledge dissemination hinder informed decision-making and the integration of all sanitation elements. This chapter describes a novel method for organizing and defining sanitation systems to facilitate informed decision-making and an integrated approach. Technologies are categorized based on their 'Product-Process' specificity and then linked into logical systems using a 'Flowstream' concept. Technologies are grouped and used to construct seven logical systems. Additionally, according to the flowstream, suitable technologies are grouped and given a score for each of the criteria. The advantages and shortcomings of the flowstream approach to sanitation system planning and the differences between 'system' and 'technology' are discussed and a set of terms and concepts that can be used to standardize the way in which sanitation is thought of and communicated about is proposed.

5.1 Introduction

Most sub-Saharan African countries are not on track to reach the sanitation target as stated in the Millennium Development Goals, which is to halve the number of people without access to adequate sanitation by the year 2015 (WHO and UNICEF 2004). Despite numerous efforts and campaigns the reality is that sanitation projects which were implemented often do not show the desired impact and have not been replicated at a large-scale across the regions of need. The marginal improvements can partly be explained by the fact that often a (centralized) water-based conventional sewer system

E. Tilley, C. Zurbrügg (✉), and C. Lüthi
Department of Water and Sanitation in Developing Countries (Sandec), Eawag: Swiss Federal Institute of Aquatic Science and Technology, Ueberlandstrasse 133, Duebendorf, Switzerland
e-mail: elizabeth.tilley@eawag.ch; christian.zurbrugg@eawag.ch; christoph.luthi@eawag.ch

B. van Vliet et al. (eds.), *Social Perspectives on the Sanitation Challenge*,
DOI 10.1007/978-90-481-3721-3_5,

is seen as the final solution to all sanitation problems in urban and peri-urban or even rural areas, irrespective of the differences in the physical and socio-economic conditions. The conviction that a conventional sewer system is the final goal of sanitation is driven by the desire to replicate the industrialized country model.

In fact, there is not much variation between the sanitation solutions in industrialized countries. Ample precipitation is evenly distributed throughout the year, and there is an existing drainage network of canals to divert rainwater and prevent floods. This naturally gave rise to sewage systems where drainage canals and flushing water were used for removing human waste materials from the inhabited areas. Consequently, increased discharges of waste water caused by the rapid growth of urban areas increasingly exceeded the assimilation and self-purifying capacity of water bodies. To reverse this effect, engineers designed and constructed sewage treatment plants; at first, simple sedimentation technologies, then biological treatment to reduce nutrient and pathogenic loads. Existing waste water treatment plants were continually upgraded and expanded with more sophisticated and expensive treatment steps. Thus, water-flush toilets connected to a water-borne sewer system and waste water treatment facility became the 'state of the art' to solve the sanitation problems in the urban areas of the North. Consequently, this was also seen as the goal for urban areas in the South. For many engineering professionals it is, even now, still considered to be the only feasible solution to Urban Environmental Sanitation.

Only recently has the notion of centralized sanitation been challenged. Research and development are more and more targeting alternative, often decentralized, approaches and solutions to the increasing environmental sanitation problem. Technical and processing innovations which facilitate the beneficial reuse of human waste products have become increasingly relevant. Urine and faeces separation and their reuse in agriculture (Ronteltap et al. 2007; Pronk et al. 2007; Tilley et al. 2008), grey water separation and reuse (Morel and Diener 2006) and increasingly (though still limited) faecal sludge collection and treatment for reuse (Cofie et al. 2006; Kengne et al. 2006; Koné et al. 2007) are but a few examples of the radical, yet practical research which is charting a new course in sanitation research. Such advances in environmental sanitation are also being discussed, developed and piloted in the North (Otterpohl et al. 2004; Jönsson and Vinnerås 2007; Rossi et al. 2009).

However, despite innovations and many publications in the sector, there is an increasing lack of focus on the integration of individual technologies into systems and no clear way of comparing technologies, systems and/or planning concepts. One of the challenges for improving sanitation in rural West-Africa is not building more toilets, but understanding the connections between the sanitation system components and what technologies and/or processes (e.g. transportation and management) are required to ensure that the whole system works in a way which is sustainable and healthy (Eales 2005).

NETSSAF 'Network for the Development of Sustainable Approaches for large scale implementation of Sanitation in Africa', a coordination action sponsored by the European Commission had, as one objective, to bridge this gap in knowledge and create synergies to support large-scale implementation of sustainable sanitation systems in peri-urban and rural areas of West Africa. By identifying and clearly illustrating sanitation systems, the different products that are generated and the corresponding technologies needed to handle those products, NETSSAF contributed to informed decision-making and advanced the discussion regarding appropriate options and alternatives.

It was recognized in the early phases of sanitation projects in developing contexts that the social, cultural and management issues are as, if not more, important than the technical design, and without considering these elements a project is likely doomed to fail. There are numerous guidance documents for participatory approaches to sanitation planning (Eawag 2005; IWA 2006; Ockelford and Reed 2002). Similarly, new works have been published which seek to alter the 'paradigm' in which sanitation systems are thought of, both in terms of the management structure and the method of implementation (Mara and Alabaster 2008) but there have been few, if any attempts to redefine the paradigm of system design. NETSSAF had the aim to highlight similarities as well as differences between the planning tools available and therefore give guidance to policy and practice. This chapter focuses on the aspects of sanitation technologies and their compatibility to enable a functioning sanitation system.

The conventional planning approach for urban environmental sanitation has been one in which planners and engineers first defined the needs of the beneficiaries and then decided what type of infrastructure and services should be provided. Sector professionals then translated hypothetical demand into project designs based on sewerage and treatment technologies commonly used in the industrial cities of Europe and the United States (Wright 1997). Such supply-driven approaches have seldom been appropriate in the developing country context as many examples illustrate. In Accra, Ghana, 20 years after construction of a sewerage system designed for 2,000 connections, only 130 connections were made. In Ma'an, Jordan, only 690 connections were established to a system designed for 6,000 connections, and in Addis Ababa, Ethiopia, during the 10 years after construction of the new sewerage system, only 10% of the expected connections were made (Wright 1997). In other cases alternative sanitation systems were planned and implemented but were not thought through from beginning to end with the consequence that key issues were overlooked which led to failure or severe environmental pollution or health risk. Faecal sludge is a typical example of a product which is generated in the middle of, or at unforeseen points in the flowstream, but often neglected in strategic planning. An example is the Strategic Sanitation Plan for Ouagadougou (PSAO) which, at a cost of US$ 14.6 million, has three main components:

- On-site sanitation for most of Ouagadougou. Community workers are paid by PSAO to encourage households to upgrade their sanitation facilities by installing one of several options for the disposal of excreta and/or soakaways for sullage disposal.
- Construction of latrines for schools and provision of educational material about hygiene and sanitation for teachers.

- A sewerage network for the city centre and industrial area and a waste water treatment facility using stabilization ponds. Industries are required to pre-treat the waste water they discharge.

For on-site sanitation facilities, households are offered a range of technical sanitation options which can be adapted to suit their financial means. They can choose between a rehabilitation of traditional latrine or the construction of a ventilated improved pit (VIP) or pour-flush latrine. For the disposal of sullage, the soakaway is the only available option. After the household has chosen a technical option, a qualified artisan is then contracted to make it. A sum of 0.16 million euro (€) helped to subsidize the components of the latrines and sanitary facilities. Households most frequently chose upgrading of traditional latrines (US$20–30) rather than VIP (US$100) given the large difference in cost. Because of limited access and costly water, only less than 1% chose pour flush systems. Although the results of the program are quite impressive, there were unfortunately, no plans or strategies for what came after the toilet facilities; that is, sludge collection or treatment facilities. This is a clear example of what happens when a systems approach is not taken, that is where a specific flowstream was overlooked and not considered from beginning to end. Excreta and faecal sludge from the various on-site facilities must be collected, transported and treated appropriately to either allow for reuse or safe disposal. Financing mechanisms for regular emptying and the assurance that the collected excreta and sludge is not dumped indiscriminately must also be addressed. The initial plans of PSAO included 28 drying beds at the waste water treatment plant but they were designed in such a way that they could neither dewater nor dry excreta, faecal sludge nor the sludge from the ponds. The lack of consideration of the faecal sludge component in the sanitation system has subsequently been realized and acknowledged and is now being rectified through a detailed study and an improved plan for sludge management. Furthermore, the households that have had a latrine rehabilitated or a soakaway constructed tend to belong to the middle class and not the urban poor, which means that the intended impact on pro-poor development is not consistently achieved. This is due to the fact that the middle class has better access to information about the subsidy scheme and on how to apply for subsidies (Vezina 2002; Savina and Kolsky 2004). This example also shows how important it is to not only consider technical issues but also social factors and interventions in the planning of sanitation.

Identifying and then mapping the flows of each product that enters into or is generated by the system, gives a more complete view of the system and the technologies required to complete the system. More often than not, a simple technology, for example a shallow sewer, is referred to as 'sanitation system' when in fact it is simply one small, yet important technology, which is part of a complete system. Indeed, a 'system' is defined by the boundaries which are defined and re-defined depending on the focus of the discussion. To eliminate ambiguity and normalize the lexicon, we propose that a sanitation system is the combination of flowstreams; each one a logical series of technologies which process each product that is introduced to, and generated within, the system. We do not propose to standardize the names of different technologies but rather, we propose a naming convention and a way of conceptualizing sanitation systems. To do so, we define the components of a system: processes, products, technologies and flowstreams. Furthermore, we propose seven unique systems (see Section 5.3) which address all phases of product generation, collection, transport, transformation and terminal use.

5.2 Concept Development and Methodology

NETSSAF is a consortium of 19 members with equal representation from the North and the South. Each relevant field of resource management in sustainable sanitation is represented which allows for the maximum integration of knowledge, expertise and experience between partners. The organization and structure of the work presented in this paper grew organically and iteratively through the input and discussions between all partners. The collaborative approach allowed for an extensive set of technology descriptions to be compiled, as well as allow for the joint, iterative development of the system diagrams to best reflect the realities of sanitation systems today.

Initially, the selected technologies and their internal groupings were chosen by the members of the consortium. The choice was determined by what was considered to be potentially feasible in more general terms for West Africa or developing countries. Each member was requested to submit a short, complete summary of the technologies that he/she was most familiar with. Concurrently, the structure of the seven different systems was defined, though without populating the system with specific details (e.g. technologies or flowstreams).

To assess the qualities of the technologies, the members defined a set of criterion: Health issues; Impact to environment; Technical characteristics; Economic and financial issues; and Social, cultural and gender issues. The assessment was conducted by flowstream, that is, all technologies that were suitable for processing a given flowstream were grouped and ranked (using the criteria developed) in order to compare 'apples and apples'. For instance: the suitability or ability of each technology to meet the desired objectives (e.g. health, social, technical objectives) in the context of the flowstream.

5.2.1 System Design

The history of sanitation engineering is long and comprehensive; the NETSSAF systems are not new in terms of the information that they contain, but rather in the way they organize and systematize pre-existing information into a standardized format and by contextualizing it in general terms for developing country situations. By thinking of a sanitation system as a matrix of products (wastes) and processes (tasks) which intersect at technologies (infrastructure and activities), a complete 'cradle to grave to cradle' system can be designed without excluding essential process steps or wastes which could be the downfall of the best-intentioned design. The elements of the systems are briefly summarized in the following sections. The descriptions are then followed by a thorough discussion of the complete systems.

5.2.2 Products

Although different human wastes have very different properties (and reuse potentials) they have, classically, been grouped together and labeled as 'sewage' or 'human wastes' (Tchobanoglous et al. 2003). In Material Flow Analysis terms (MFA), 'goods' are defined as 'substances or mixtures of several substances with functions valued by men' (Baccini and Brunner 1991). These two definitions have connotations at the extreme ends of the 'good' and 'bad' spectrum; we propose instead to use the term 'product' as a more neutral, and in fact, accurate term to describe and include all organic and inorganic materials that flow through a sanitation system. Each product differs in its characteristics due to mixing or separating different types of excreta and quantities and types of water. By carefully differentiating between products, and therefore the inherent characteristics, the feasible and subsequently most appropriate technologies and systems can be selected. Although there are nearly infinite combinations, we define eight unique products which would be considered in the systems design:

- **Urine** is the liquid waste excreted by the body.
- **Faeces** are the (semi-)solid wastes excreted by the body.
- **Excreta** is the combination of urine and faeces.
- **Black water** is the mixture of excreta and flushing water, along with anal cleansing water or dry cleansing material (depending on what is practiced).
- **Faecal Sludge** is the general term for the undigested or partially digested slurry or solid that results from the storage or treatment of blackwater or excreta.
- **Beige water** is water that is used for anal cleansing after defecation.
- **Grey water** is water that has been used for bathing, hand-washing, cooking, clothes-washing or other types of cleaning.
- **Stormwater** is the general term for the rainfall that runs off from roofs, roads and other surfaces before flowing towards low-lying land. It is the portion of rainfall that does not infiltrate into the soil.

The greatest benefits can be achieved through product separation and product targeted processing. Product separation minimizes pathogen spread, concentrates nutrients and facilitates beneficial reuse (or at least benign disposal). For instance, human urine collected during 1 year (ca. 500 L), depending on diet, contains 2–4 kg nitrogen, and with the exception of some rare cases is sterile when it leaves the body (Tilley et al. 2008). Another example is grey water which may contain few pathogens and its flow of nitrogen is only 10–20% of that in blackwater. However, it accounts for approximately 60% of the waste water produced in households with a blackwater flowstream (Morel and Diener 2006).

These eight products are those that, separated or combined, are unique enough to necessitate or allow for product-specific technologies; a crucial deviation from the black box approach to sewage treatment.

5.2.3 Process

A 'process' is a task; a process contains, transforms, or transports products to other process steps or a final point of use or disposal. In this document we refer to six different processes. Technologies which perform the same process are grouped together under the same heading; they are differentiated not by the task that they perform, but rather by the products they treat and their connectivity to other technologies and processes. The processes are defined below.

- **User interface** describes the way in which users of a sanitation facility access and interact with the sanitation system.
- **On-site Collection, Storage and Treatment** describes the technologies that can be used onsite (i.e. at the household/compound level) to collect, store and (partially) treat different products.
- **Transport** describes the way in which products are transferred from one process to another. It is important to note here, that although there are many points in a system where products must be transported, we limit the definition in this context to be between On-Site Collection, Storage and Treatment and Treatment offsite, since this is often, the most lengthy and costly transport step.
- **Treatment offsite** describes the technologies used to reduce pathogens and/or nutrient loads of the products.
- **Reuse** describes the technologies and/or methods which allow some benefit to be derived from a flowstream.
- **Disposal** describes the technologies and/or methods which allow the (transformed) products to be returned to the environment in a benign/non-detrimental way.

The definition and organization of technologies into process categories is one of the most important aspects of the NETSSAF work; too often two or more technologies are combined and referred to as a single entity, even though each part performs a distinctly unique task. Most commonly, toilets and pits are referred to as a 'latrine' (e.g. 'pour-flush latrine') when the actual processes performed by said 'latrine' are left unclear. Similarly, a septic tank is often assumed to come part and parcel with a leach-field, despite the fact that many septic tanks are not connected to a leach-field, and are often better suited to a small-bore sewer connection. By eliminating the ambiguity of what 'process' a technology is responsible for, a more concise parlance and design concept will result.

5.2.4 Technology

A technology performs a product-specific process, that is, it occurs at the product-process intersection. A technology is a product-specific method or tool designed to collect, store, transform (change), move, or dissipate a product. A comprehensive list of common, alternative and experimental technologies were compiled, described and organized according to process and flowstream. The information compiled is a broad-reaching and diverse summary with topics ranging from micro-flush toilets to septic tanks to excreta treatment with urea.

5.2.5 Flowstream

In the MFA sector, 'flows' are vectors, that is they have a unit and a direction (Montangero and Belevi 2007). A flow would be, for example, 100 t of phosphorus per year applied to a field. This concept of moving mass from one process to another was adapted here to describe the entire path, for example lifecycle, that a product or mixture of products takes from the point of generation to the ultimate point of reuse or disposal. It is the entire set of changes (chemical, physical, biological, spatial) that a product undergoes from the point of generation to the point of dissipation into the environment. The flowstream is the sum of technologies for a product (or group of products) which flow through the process steps.

However, rarely does a single product flow through the process steps alone; generally two or more products will flow, and be processed concurrently. Combining products may be a function of the User Interface (i.e. how people generate products), because of volume (i.e. one product does not warrant its own technology) or for other economic or logistical reasons. We define 11 different flowstreams; the difference between the number of products and the number of flowstreams is due to the fact that some combinations of products require their own flowstream. The flowstreams are summarized in Table 5.1.

These 11 flowstreams vastly amplify the specificity of a waste's definition, in terms of volume, pathogenicity, nutrient load, and reuse potential. By clearly defining and understanding the actual flowstreams that are being created and in need of processing, a sanitation system can be more accurately assessed, improved and/or designed.

5.2.6 Sanitation System

So the flowstream is the sequence and sum of the product-specific technologies, the sanitation system is the sum of the flowstreams. By considering a system, that is, all of the products that are generated, the way they are separated and/or combined, the processes they pass through, the technologies that perform the processing task, and

Table 5.1 Flowstream composition and description

Flowstream	Description
Blackwater	Products: urine, faeces, flushing water, cleaning material or beigewater. Description: Lack of grey water in this flowstream may limit the self-cleansing velocity in a sewer network given the reduced liquid content.
Grey water	Products: grey water Description: Grey water accounts for 50–80% of the outflow produced at household level, although this very much depends on local conditions. It contains few, if any, pathogens and 90% less nitrogen than blackwater and therefore does not require the same treatment processes as blackwater or mixed waste water.
Faecal sludge	Products: faecal sludge Description: Faecal sludge can be mostly biological (e.g. from trickling filters) or mostly raw faecal material (e.g. from pit latrines). We do not distinguish between faecal sludge and biosolids[a]
Brownwater	Products: faeces, flushing water, cleaning material and/or beigewater Description: Brownwater results from wet-urine diversion systems. Typically, brown water is transported through sewers and is treated offsite. It is similar to blackwater, however with the urine removed, the nutrient levels are significantly lower.
Urine flowstream	Products: urine Description: Urine is collected by a urine-diverting user interface. Separately collected urine from a healthy person does not contain pathogens. However, urine may still be contaminated easily by traces of faeces.
Excreta flowstream	Products: urine and faeces Description: The excreta flowstream is collected with a dry user interface, i.e. without flushing water. Given the low liquid content (and therefore, reduced ease of transport), treatment occurs on-site.
Faeces	Products: faeces Description: Faeces are collected parallel to urine in a urine-diverting user interface. This flowstream resembles the excreta flowstream but is drier (as urine is missing) and is therefore transport limited. Treatment occurs on-site.
Beigewater	Products: Beigewater (anal cleansing water) Description: Beigewater, although very dilute, contains a significant amount of faecal material and is therefore pathogenic and should be treated appropriately.
Mixed blackwater and grey water	Products: urine, faeces, flushing water, and grey water (stormwater may or may not be diverted into the sewer and mixed with this flowstream), and cleaning material or beigewater. Description: This is the most common flowstream in industrialized city systems and is commonly referred to as 'waste water'.
Brownwater mixed with grey water	Products: faeces, flushing water, grey water, and cleaning material or beigewater. Description: This flowstream is similar to the blackwater mixed with grey water flowstream except that the urine has been separated out. With the separation of urine, brownwater contains lower concentrations of nutrients (as nitrogen and phosphorous is mainly contained in the urine). This aspect of low nutrient concentrations is further enhanced by the inclusion of grey water, which further decreases the nutrient concentrations. This flowstream is rare, as it is dependent on water-based urine-diverting toilets, which have not been widely installed.

(continued)

Table 5.1 (continued)

Flowstream	Description
Excreta mixed with grey water	Products: urine, faeces, grey water (cleansing material or beigewater) Description: This is a commonly seen in the developing world, though it is not generally recommended. In dense areas with a scarcity of space, on-site sanitation technologies (e.g. pit latrines) are used for both excreta and household waste water disposal.

[a]In strict engineering terms, biosolids are defined as sludge or solids that are treated and/or stabilized (Tchobanoglous et al. 2003)

the products that are the result of the process, the boundaries and components of the system can be understood.

A system however, is not a simple combination of products and technologies which can be chosen at will; technologies must be linked logically, products must be combined/ separated appropriately, and process steps ordered to reflect the latter.

5.2.7 System Diagrams

To visualize the 'system' concept, system diagrams were developed (Fig. 5.1). The system is visualized as a matrix: process titles are written across the top (column headings); products are written down the left (row headings). Where the products and processes intersect, there is a box indicating a technology. Seven different systems were developed; four which were classified as 'wet' and three which were classified as 'dry'. The words 'wet' and 'dry' do not necessarily describe the humidity of the material that is produced, but rather, the presence or absence of water used for transportation. Arrows indicate the flow of products, move from left to right and indicate when there are two paths that the product can take depending on the desired outcome. Flowstreams are separated by dashed lines and are read from right to left; the flowstream title is written down the right side of the diagram. Flowstreams are read from right to left because the flowstream begins at the point where the different products are combined into the single flowstream that will flow through the process steps as a single phase; sometimes the flowstream will flow through five process steps, sometimes just through one. A summary of the seven systems is presented in Table 5.2 which makes it clear that each system is comprised of flowstreams, oftentimes more than one.

Figure 5.1 is a visual representation of System F: *Dry urine, faeces and grey water diversion system* (a dry, or waterless, system). This system is characterized by the separation of urine, faeces and grey water – at the source – into three different flowstreams, and where anal cleansing water is used, a fourth flowstream. In this way, each flowstream can be more appropriately managed in terms of its volumetric flow, nutrient and pathogen content and handling characteristics. This diversion facilitates more targeted treatment and end use for the different fractions. This system requires a urine-diverting user interface. Urine is collected through the front

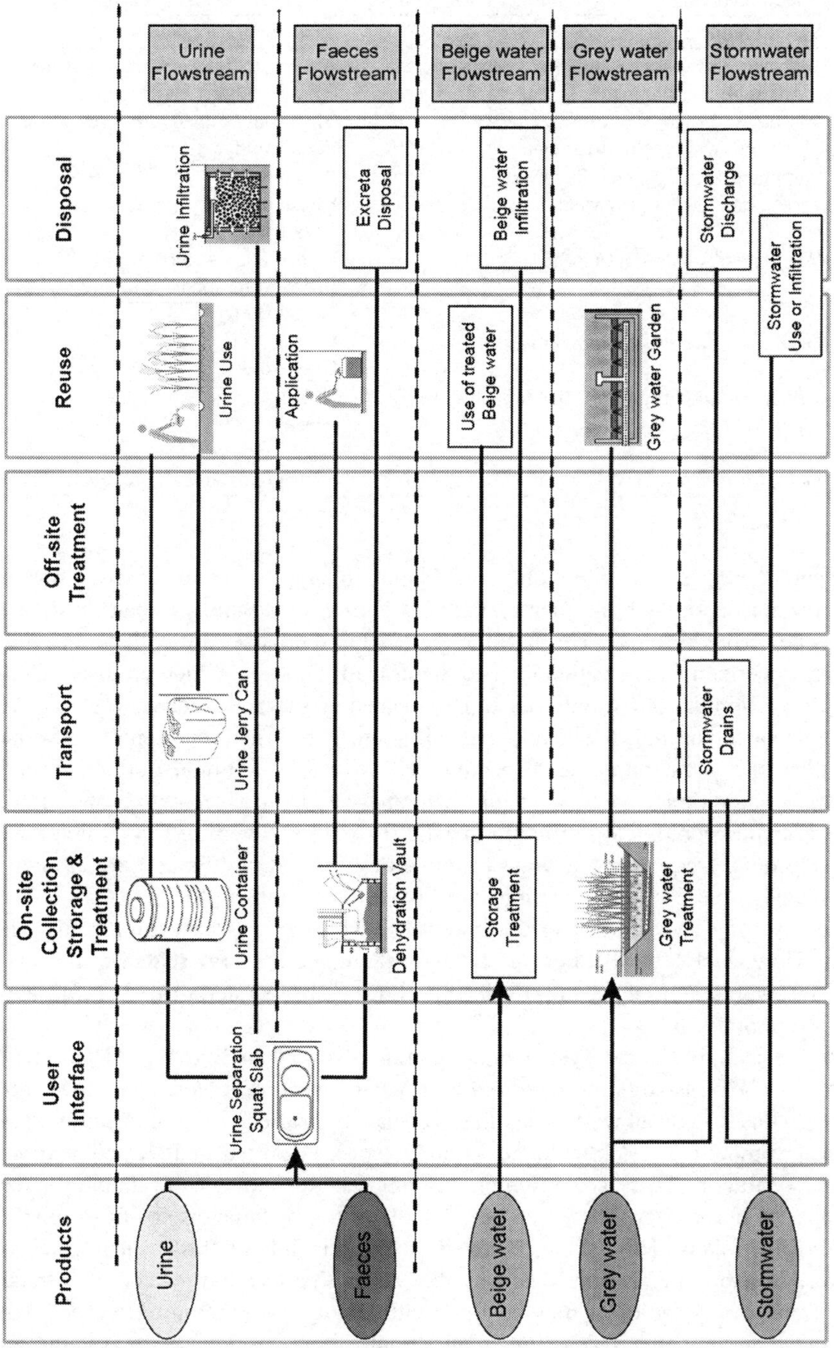

Fig. 5.1 System F: dry urine, faeces and greywater diversion system

Table 5.2 Seven sanitation systems as characterized by the NETSSAF project. The first four are 'wet' systems which involving flushing of human waste with water; the other three systems are 'dry'

	System name	Flowstreams included
A	Wet mixed blackwater and grey water system with offsite treatment	Blackwater mixed with grey water Faecal sludge
B	Wet mixed blackwater and grey water system with onsite treatment	Blackwater mixed with grey water Faecal sludge
C	Wet blackwater systems (blackwater separated from grey water)	Blackwater Faecal sludge Grey water
D	Wet urine-diversion system	Urine/yellowwater Brownwater mixed with grey water Faecal sludge
E	Dry grey water-separate system	Excreta Grey water
F	Dry urine- and grey water-diversion system	Urine Faeces Grey water
G	Dry all mixed systems	Excreta mixed with grey water

outlet and conveyed to a collection vessel (a tank in larger, more expensive systems or a jerry can in smaller, simpler systems), a garden or possibly a soak pit, if the urine is not brought to use. Through the rear outlet, the faeces are collected in two dehydration containers or vaults located underneath the toilet which are used alternatly for a minimum of 6 months each. Dry cleansing material (such as toilet paper) can be dropped through the rear outlet, although it is often kept separate. Some urine-diverting squat pans are also equipped with an additional outlet for anal cleansing water (beige water), which is then treated, in a separate flowstream. These so-called ecological sanitation or UDDTs (Urine Diverting Dry Toilets) are becoming more frequent as they are promoted by bi- and multilateral donor agencies as well as research organizations worldwide (Sanimap 2008).

The case of eThekwini District, Durban, South Africa is an example for this system. Here, such UDDT toilets are being implemented at large scale for the peri-urban unsewered area of the city (Macleod 2005) and separated urine is infiltrated using a soakpit.

Another example is the system most commonly used in Europe and Northern America, the *Wet mixed blackwater and grey water system with offsite treatment* (not shown). In this system, all waste water that is created by households, institutions, industries and commercial establishments is collected, transported and treated without product separation. There are different user interface technologies available for the collection of blackwater. These can be by high- or low-volume cistern-flush toilets, or pour-flush toilets. This system is based on the principle of flush water which is used as a transport media and to convey the wastes to a location for common treatment. Transport technologies may be pipes with gravity flow, pressure flow, or using vacuum technology. In a developing country context, gravity flow sewer systems are often used in very densely populated city centres, however they are often lacking a well functioning treatment plant. One interesting example is the sewerage network of Ouagadougou which covers part of the city centre and industrial area and leads to waste water stabilization ponds for treatment (Vezina 2002). A variation of this conventional sewerage network system is the simplified sewer, also called condominial sewer which is common in Brazil and other parts of Latin America (Melo 2006). Simplified Sewers are constructed using smaller diameter pipes laid at a shallower depth and at a flatter gradient than conventional sewers. The Simplified Sewer allows for a more flexible design associated with lower costs and a higher number of connected households. Expensive manholes are replaced with simple inspection chambers. Another key design feature is that the sewers are laid within the property boundaries, rather than beneath the central road. These sewers are also often referred to as condominial sewers because the community will purchase, and connect to, a single legal main sewer connection to which the combined effluent of the condominial sewer network flows. Generally there is no special consideration to resource recovery although in an industrialized context the sludge generated at the waste water treatment plant is often used for energy generation or reuse in agriculture.

5.3 Assessment

As an additional tool to complement the technology descriptions and the system diagrams, each technology was evaluated using a qualitative scale which is summarized in Table 5.3.

Technologies were grouped according to the flowstream for which they were most suitable and then given a score for each of the criterion within the following categories: Health issues; Impact to environment; Technical characteristics; Economic and financial issues; and Social, cultural and gender issues. The assessment was applied directly for the flowstream, that is, a technology could receive the highest possible score for a criterion in the context of one flowstream and the lowest possible score for the same criterion in the context of a different flowstream. Not only does this arrangement of flowstream-based comparison allow the user to compare technologies for each criterion listed, but it indicates what technologies are appropriate for each flowstream. Table 5.4 is the ranking table developed for the urine flowstream.

Table 5.3 Qualitative descriptors for assessment of technologies

Qualitative Descriptor	Meaning
+ +	The technology fulfilled this criterion very well
+	The technology fulfills this criterion
0	The technology is neutral to this criterion
-	The technology does not fulfill this criterion well
- -	The technology does not fulfill this criterion at all
Shaded	The criterion is not applicable for this technology

Table 5.4 Ranking chart developed for the urine flowstream

Technologies applicable for: Urine flowstream		User Interface			Onsite C,S,T	Transport			Offsite Treatment		Reuse		Disposal
		Urine diverting toilets wet	Urine diverting toilets dry	Urinal	Storage tank	Urine pipes	Manual urine transport	Truck for urine transport	Off-site urine storage tank	Struvite precipitation	On-site application	Off-site application	Soakaway pit
Health issues													
Reduces exposure	Of users	0	0	0	0	++	–	++	0	0			0
	Of waste workers				0	+	––	–	0	0	0	0	
	Of reusers				+				+	–	0	–	
	Of "downstream" population				+				+	+	0	0	0
Hygienization					+				+	+			0
Increases health benefits					0				0	+	++	+	0
Impact to environment/nature													
Use of natural resources	Needs low land requirements				0	0			0	–	–	–	+
	Needs low energy requirements	++	++	++	++	++	+	––	++	––	++	–	+
	Uses mostly local material	–	0	+	+	–		––	+	––	+	–	+
	Low water amounts required	–	++	+	+				+	+	+	+	+
Low emissions and impact to the environment	Surface water				++			–	++	++	+	+	–
	Ground water				++			–	++	++	+	+	–
	Soil/land				+			–	++	++	0	0	–
	Air				++			–	++	++	0	0	0
	Noise, smell,	0	0	0	0	0	0	–	0	0	–	–	+

Good possibilities for recovering resources	Nutrients	++	++	++	++				++	++	++	++	0
	Energy	0	0	0	0				0	–	0	0	0
	organic matter	++	++	0	0				0	0	0	0	0
	Water	+	+	+	+				+	0	0	0	0
Technical characteristics													
Allows simple construction and low level of technical skills required for construction		–	–	0	+	–			+	––	++	0	++
Has high robustness and long lifetime/high durability		–	0	0	+	0	+	–	+	––	++	–	++
Enables simple and low cost O&M and low skills required		–	0	0	+	+	+	–	+	––	++	–	+
Economical and financial issues													
Has low construction costs (unit cost per household)		–	0	+	–	––	+	–	–	––	++	–	+
Provides benefits to the local economy (business opportunities, local employment, etc.)		+	+	+	0	+	+	+	0	+	+	+	+
Has low operation and maintenance costs		–	0	+	+	–	++	––	–	–	++	–	++
Provides benefits or income generation from reuse		++	++	++	++				++	++	++	++	––
Social, cultural and gender													
Delivers high convenience and high level of privacy		++	++	++	0								++
Requires low level of awareness and information to assure success of technology		––	––	+	+	++	–	0	+	––	–	–	+
Requires low participation and little involvement by the users		–	–	+	0	++	–	–	0	0	0	0	+
Takes special consideration for issues of women, children and elderly		0	0	0									

The ranking chart shown here is one of the charts developed with the fewest technologies selected; other flowstreams such as blackwater have a larger number of technologies for which they might be applicable. In the case of the urine flowstream, there are technologies indicated for each of the processes, although in some cases, for example Onsite Collection, Storage and Treatment, there is only one technology shown and ranked. Without a doubt, there are infinite technology variations that could be used for different applications under different circumstances, however for this chapter we limit the technologies in the ranking chart to those that are summarized and described in the output of the NETSSAF project. Note also that all criteria could not be applied to each technology. For example 'urine pipes' are not able to take special consideration of issues related to women, children and the elderly and for this reason is shaded grey (it is important that the criteria are read literally for the single technology and are not projected onto the system as a whole).

These assessment charts allow for the rapid identification and assessment of technologies on a flowstream-by-flowstream basis and also between flowstreams. They can be used iteratively with the system diagrams to populate the general technology spaces with specific technology choices.

5.4 Conclusions

By developing a standardized system for identifying, categorizing and linking the technologies of a sanitation system, we have attempted to reduce ambiguity and allow for greater transparency and completeness within sanitation planning. The seven unique systems defined contain multiple combinations which, when all considered, represent a nearly complete set of feasible sanitation systems. Most importantly, the system diagrams encourage the consideration and inclusion of all products. With the ability to identify overlaps and gaps, redundancies and deficiencies can be reduced.

The qualitative rating system developed allows for the rapid identification and assessment of technologies for specific processes and between flowstreams. Although the ranking scale is qualitative, and not exhaustive, it is a useful tool for rapidly assessing the relative adequacy of a given technology for a given flowstream. As well, the system diagrams can be populated by elucidating the most appropriate technologies from the ranking charts.

Above all, the work described here (Zurbrügg and Tilley 2007) should not be seen as a stand-alone document. It is our intention that it will be used as policy tool in the context of a comprehensive implementation strategy such as proposed by NETSSAF, HCES (Eawag 2005), Sanitation 21 (IWA 2006), PROVIDE or a similarly guided project. The systems approach and an overview of technologies has been developed further by Tilley et al. (2008) and serves to facilitate decision-making during the planning process. The systems approach as we described it should not be used as justification for a U-turn towards top-down planning of sanitation. Rather, it should fill the need for a system design tool within a participatory planning process. It is our sincerest hope that by encouraging and fostering a common language of technical communication, more successful sanitation projects will be realized, with the ultimate goal of generating increased motivation, innovation, and ultimately improved health for all those who still require adequate sanitation.

Acknowledgements The authors would like to thank the entire NETSSAF consortium (www.netssaf.net) for their dedicated and supportive collaboration in this, and all other activities in the project. Additionally the authors would like to thank the European Union for funding this project under the EU-FP6 Framework. Finally the authors would like to acknowledge the support of: Eawag, the Swiss Federal Institute of Aquatic Science and Technology and the Swiss National Centre of Competence in Research (NCCR) North-South: Research Partnerships for Mitigating Syndromes of Global Change.

References

Baccini, P. & Brunner, P. H. (1991). *Metabolism of the anthroposphere*. Heidelberg/New York: Springer.

Cofie, O., Agbottah, S., Strauss, M., Esseku, H., Montangero, A., Awuah, E., et al. (2006). Solid-liquid separation of faecal sludge using drying beds in Ghana: Implications for nutrient recycling in urban agriculture. *Water Research, 40*(1), 75–82.

Eales, K. (2005). *Bringing pit emptying out of the darkness: A comparison of approaches in Durban, South Africa, and Kibera, Kenya*. Sanitation Partnerships Series, Building Partnerships for Development in Water and Sanitation, BPD Water and Sanitation, London, UK.

Eawag. (2005). *Household-centred environmental sanitation: Implementing the Bellagio principles in urban environmental sanitation (Provisional Guideline for Decision-makers)*. Duebendorf, Switzerland: Eawag (Swiss Federal Institute of Aquatic Science and Technology).

IWA. (2006). *Sanitation 21: Simple approaches to complex sanitation - a draft framework for analysis*. London: International Water Association. Retrieved March 27, 2009, from http://www.iwahq.org/uploads/iwa%20hq/website%20files/task%20forces/sanitation%2021/Sanitation21v2.pdf

Jönsson, H., & Vinnerås, B. (2007). Experiences and suggestions for collection systems for source-separated urine and faeces. *Water Science and Technology, 56*(5), 71–76.

Kengne, I. M., Amougou, A., Bemmo, N., Strauss, M., Troesch, S., Ntep, F., et al. (2006). Potentials of sludge drying beds vegetated with Cyperus papyrus L. and Echinochloa pyramidalis (Lam.) Hitchc. & Chase for faecal sludge treatment in tropical regions. *Proceedings of the international conference on wetlands systems for water pollution control* (Vol. 2, pp. 943–953). Lisbon, Portugal.

Koné, D., Cofie, O., Zurbrugg, C., Gallizzia, K., Moser, D., Drescher, S., et al. (2007). Helminth eggs inactivation efficiency by faecal sludge dewatering and co-composting in tropical climates. *Water Research, 41*(19), 4397–4402.

Macleod, N. A. (2005). The provision of sustainable sanitation services to peri-urban and rural communities in the eThekwini municipality. *3rd International Ecological Sanitation Conference* (pp. 47–51). Durban, South Africa.

Mara, D. & Alabaster, G. (2008). A new paradigm for low-cost urban water supplies and sanitation in developing countries. *Water Policy, 10*(2), 119–129.

Melo, J. C. (2006). *The experience of condominial waterand sewerage systems in Brazil: Case studies from Brasilia, Salvador and Parauapebas*. Water and Sanitation Program, Latin America (WSP-LAC). Retrieved March 27, 2009, from http://www.wsp.org/UserFiles/file/BrasilFinal2.pdf

Montangero, A. & Belevi, H. (2007). Assessing nutrient flows in septic tanks by eliciting expert judgement: A promising method in the context of developing countries. *Water Research, 41*(5), 1052–1064.

Morel, A. & Diener, S. (2006). *Greywater management in low and middle-income countries, review of different treatment systems for households or neighbourhoods*. Dübendorf, Switzerland: Swiss Federal Institute of Aquatic Science and Technology (Eawag).

Ockelford, J. & Reed, B. (2002). *Participatory planning for integrated rural water supply& sanitation*. UK: WEDC, Loughborough University.

Otterpohl, R., Braun, U., & Oldenburg, M. (2004). Innovative technologies for decentralised water-, wastewater and biowaste management in urban and peri-urban areas. *Water Science and Technology, 48*(11–12), 23–32.

Pronk, W., Zuleeg, S., Lienert, S., Escher, B., Koller, M., Berner, A., et al. (2007). Pilot experiments with electrodialysis and ozonation for the production of a fertilizer from urine. *Water Science and Technology, 56*(5), 219–227.

Ronteltap, M., Maurer, M., & Gujer, W. (2007). Struvite precipitation thermodynamics in source-separated urine. *Water Research, 41*(5), 977–984.

Rossi, L., Lienert, J., & Larsen, T. A. (2009). Real-life efficiency of urine source separation. *Journal of Environmental Management, 90*(5), 1909–1917.

Sanimap. (2008). *World Sanitation Project Map. Sanimap – a communication tool to solve sanitation problems*. Retrieved March 27, 2009, from http://www.sanimap.net/xoops2/modules/gnavi/

Savina, A., & Kolsky, P. (2004). *Mobilizing resources for sanitation*. Water and Sanitation Program – Africa Region, Field Note. World Bank, Nairobi, Kenya.

Tchobanoglous, G., Burton, F. L., & Stensel, H. D. (2003). *Wastewater engineering: Treatment and reuse*. Boston, MA: McGraw-Hill.

Tilley, E., Atwater, J., & Mavinic, D. (2008). Recovery of struvite from stored human urine. *Environmental Technology, 29*(7), 797–806.

Tilley, E., Lüthi, C., Morel, A., Zurbrügg, C., & Schertenleib, R. (2008). *Compendium of sanitation systems and technologies*. Dübendorf, Switzerland: Eawag.

Vezina, M. (2002). *The Ougadougou strategic sanitation plan: An holistic approach to a city's problems*. Water and Sanitation Program – Africa Region, Field Note 10. World Bank, Nairobi, Kenya.

WHO, UNICEF. (2004). *Meeting the MDG drinking water and sanitation target- a mid-term assessment of progress*. Geneva, Switzerland: World Health Organization.

Wright, A. M. (1997). *Toward a strategic sanitation approach: Improving the sustainability of urban sanitation in developing countries*. Washington, DC: UNDP-World Bank Water and Sanitation Program.

Zurbrügg, C., & Tilley, E. (Eds.). (2007). *Evaluation of existing low-cost conventional as well as innovative sanitation system and technologies*. Workpackage 3 – Assessment of Sanitation Systems and Technologies. NETSSAF, Deliverable 22 & 23. Project no. 037099.

Chapter 6
A Learning and Decision Methodology for Drainage and Sanitation Improvement in Developing Cities

Joost van Buuren and Astrid Hendriksen

Abstract In urban areas in the South drainage and sanitation systems in various states of development coexist. At a certain stage authorities have to take decisions concerning the transition from on-site to off-site systems or the unification (centralization) of the decentralized systems in their constituencies. A method is presented to support decisions about drainage and sanitation systems based on multi-criteria decision analysis in combination with stakeholder dialogues. This participatory methodology brings about a learning process in which experts and non-experts are enabled to connect local experience with systemic knowledge, in order to generate, assess and select sustainable drainage and sanitation solutions. The method is supported by a database which describes 58 drainage and sanitation options, assessment objectives, a screening aid and a performance matrix. The options are constructed on the basis of different water-using and reuse-oriented toilet types, the use/non-use of septic tanks, and different collection systems for stormwater and domestic waste water, which distinguish themselves by different environmental performance and the way the unwanted water is transported. Through application during workshops in Ho Chi Minh City (Vietnam) and Kampala (Uganda) insight was gained in the possibilities of the method and in items to be improved.

6.1 Introduction

As settlements in the fringes of cities in the South develop from an originally rural condition to densely populated urbanizations, drainage and sanitation systems undergo incremental changes. For example: at low building densities and water

J. van Buuren (✉)
Sub-department of Environmental Technology, Wageningen University, Bomenweg 2
6703 HD Wageningen, The Netherlands
e-mail: joost.vanbuuren@wur.nl

A. Hendriksen
Environmental Policy Group, Wageningen University, Hollandseweg 1
6706 KN Wageningen, The Netherlands
e-mail: astrid.hendriksen@wur.nl

B. van Vliet et al. (eds.), *Social Perspectives on the Sanitation Challenge*,
DOI 10.1007/978-90-481-3721-3_6,

consumption households with pour-flush toilets and water-saving household appliances originally discharge directly to the soil near the house via a soakage pit, and storm-water is running off via natural ditches to larger surface water. Once water consumption and housing density increase, the drainage and sanitation in a certain area could develop to a system in which domestic waste water is pre-treated in a septic tank and the effluent discharged to a covered storm-water canal. The characteristics of drainage and sanitation systems are strongly determined by the living standards and the degree of urban planning applied. As in recently grown city districts in developing countries often planned and unplanned areas coexist, so do different types of drainage and sanitation systems. At a certain stage city and district authorities are confronted with choices about the improvement of drainage and sanitation. To what degree can household on-site disposal be maintained? Should communal systems be introduced? Or should large-scale sewerage be extended to new urbanizations? What type of sewer system should be opted for? Is treatment of the collected waste water affordable and what method should be chosen?

Often the decisions about drainage and sanitation systems are taken by provider-related experts who search for more options than what is already well-known, without having detailed knowledge of the situation under study and the wishes of users and re-users of products. Here, the expert dilemma is felt: knowing solutions without knowing the problem. In practice this usually means that high-flush toilets are taken for granted and existing combined sewer systems extended with as a result problems of treatment of a quantitatively and qualitatively strongly varying influent and a strong environmental impact of combined-sewer overflows (CSO).

In contrast to such an expert-driven approach, the participatory MCDA method presented in this chapter works with the following principles:

- An integrated material chain approach. Drainage and sanitation systems are material chains which should be assessed in their totality. A drainage and sanitation chain includes water supply, household appliances, on-site pre-treatment, transport, off-site treatment and reuse.
- A multi stakeholder involvement approach. Stakeholder participation is important to create commitment for the solutions chosen, especially where reuse-oriented systems are introduced.

In a *workshop-setting* relevant stakeholders advise or decide together about the most appropriate infrastructure system. The essence of the method is participatory design and evaluation of the appropriateness of several feasible options using multi-criteria decision analysis (MCDA). Apart from the local expert and non-expert stakeholders a third party in the workshop is the facilitator who may bring in expert knowledge based on the MCDA data base. Below the method is explained in more detail, first introducing the principles of MCDA and the role of the stakeholders in the process, and then the MCDA tool for drainage and sanitation system selection. The explanation of the tool is subdivided into sections about objectives, drainage and sanitation system options, the assembly of systems from building blocks and the screening and comparison of options. The chapter is completed with the outcomes of workshops in Ho Chi Minh City (Vietnam) and Kampala (Uganda).

6.2 Multi-Criteria Decision Analysis (MCDA)

In the process of solving complicated problems such as the improvement of drainage and sanitation the following consecutive stages may be distinguished: (1) problem identification, (2) diagnosis, (3) design of solution options, (4) selection of best solution, (5) intervention, and (6) evaluation. These six stages constitute the intervention cycle. After the intervention the problem should have been (partially) solved. Multi-criteria decision analysis is applied here to help stakeholders in finding sustainable and preferred sanitation options by selecting them out of a collection of feasible options, therefore it is a tool for the stages (3) and (4): the design of options and selection of the best one. In decision-making about complex interventions, such as the selection of drainage and sanitation systems, the method is used with involvement of all stakeholders relevant to the decision. The main strength of MCDA involving relevant stakeholders is the mutual learning between experts and interest groups regarding views, objectives and alternatives (Lahdelma et al. 2000). Hammond and others describe a MCDA decision-making method as consisting of five steps with the acronym PrOACT: (a) problem reconnaissance, (b) formulation of objectives, (c) generation of alternatives, (d) consequences, and (e) trade-off of strengths and weaknesses to find the most preferred solution. (Hammond et al. 1999). Problem reconnaissance in MCDA is not the analysis of the drainage and sanitation problems in a community, but rather clarifies the conditions under which solutions have to function. In our case MCDA is used as a tool to select in a participatory way the best option from a range of possible options. First, the role of the stakeholders is discussed.

6.3 Stakeholder Roles in Design of Sanitation System Options

Stakeholders are those with an interest in the planning and decision process (Lahdelma et al. 2000). The stakeholder group in sanitation-system selection can be subdivided along the dimensions experts/non-experts and professionals (paid)/non-professionals (not paid). Decision-makers will in general not participate personally in the MCDA process about drainage and sanitation infrastructure, but use its outcomes as a basis for the decision about an aimed intervention. They may want to give input to and supervise the process and interpret its results according to the requirements of the political conditions. Decision-makers are considered as non-experts. In Fig. 6.1 the roles of experts and non-experts in the five stages of MCDA are attributed according to the specific strengths of the two groups. Experts know the conditions required for technical systems to work (system options and their consequences), while non-experts have a direct interest in and detailed knowledge of the situation for which a solution is sought (problem). Both groups have a role to play in the formulation of the objectives, the consequences and the trade-off and often the non-experts also have valuable ideas about alternatives (technical options).

The invitation of participants to the MCDA process should be done on the basis of a stakeholder analysis prepared by the initiator of the MCDA process. For several

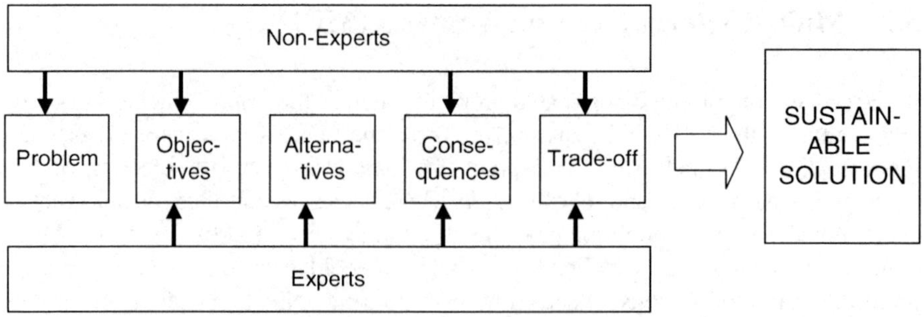

Fig. 6.1 The role of expert and non-expert stakeholders in MCDA. (→= stage in which stakeholder group plays important role)

reasons it can be difficult in practice to obtain the commitment and involvement of all stakeholders that seem relevant as per the stakeholder analysis.

Since the value of the MCDA outcome is decisively determined by the quality of the information applied in the setting of objectives, development of alternatives and assessment of consequences, the MCDA tool used in these stages is discussed in further detail below.

6.4 Objectives in Sanitation Systems Improvement

The overarching objectives of drainage and sanitation interventions are improvements in the domain of public health and environment. Technological systems applied should satisfy these two objectives, while at the same time they should also be technically functional, socially manageable and economically desirable. Jointly these five constitute the primary objectives of drainage and sanitation infrastructure. Some authors consider improved health an implicit part of improved environmental conditions (e.g. Van der Vleuten-Balkema 2003) and apply four primary objectives. As attainment of the primary objectives may be determined by many aspects, secondary objectives and tertiary objectives are defined which enable an accurate assessment of options.

Our MCDA tool proposes the use of the objectives presented in Table 6.1. This table has been composed on the basis of objectives used in the domains of water supply (Foxon et al. 2002), sanitation (Loetscher 1999; PHSSDA 2007; Zurbruegg and Tilley 2007), waste water treatment (Seghezzo 2004: 132) and urban water (Van der Vleuten-Balkema 2003).

Among objectives a distinction is made between those used for *screening*, that is distinguishing between feasible and unfeasible technological options, and objectives that are used for *comparison* of options. It is presupposed that local infrastructure, physical conditions and policies determine the choice of drainage

Table 6.1 Technology-specific objectives for assessment of sustainability of drainage and sanitation systems

Primary objectives	Secondary objectives	Tertiary objectives	Nr
Technical functionality	Compatibility with local infrastructure and physical conditions		1
	Compliance with reigning policy framework		2
	Reliability	Low level of skills needed in construction	3
		Low level of skills needed in operation	4
		Low sensitivity to irregular maintenance	5
		Independence of external supplies (e.g. power, chemicals) and services	6
	Flexibility	Ease of adaptation to fluctuations and new requirements	7
Protection of health	Avoiding exposure	Prevention of exposure of users	8
		Prevention of exposure of waste workers	9
		Prevention of exposure during reuse	10
		Prevention of exposure of downstream population	11
Environmental protection and material resources conservation	Prevention of emissions to water	Low COD emissions due to combined sewer overflows and untreated stormwater runoff	12
		Low N and P emissions due to combined sewer overflows and untreated stormwater	13
	Prevention of emissions to air	Low methane emissions	14
		Low malodours and Insects nuisance	15
	Prevention of emissions to soil	Low emissions to soil and groundwater	16
	Resource recovery	Water	17
		Energy	18
		Nutrients	19
Social manageability	Low requirements to management capacity of providers	Low requirements to institutional support and cooperation through the chain	20
		Low requirements with respect to end-user awareness	21
	High user convenience	High convenience and cultural acceptability	22
		Consideration to issues of women, children and elderly	23
Economic desirability	Economic efficiency	Low life cycle costs, total annual costs per household	24
		High benefits from reuse	25

and sanitation systems in the first place. In Table 6.1 the first two (secondary/tertiary) objectives, *compatibility with local infrastructure and physical conditions* (1) *and compliance with reigning policy framework* (2) are considered *screening* objectives. Technological systems that do not satisfy the requirements of objectives (1) and (2) are deemed unfeasible and can be omitted from the selection of feasible options. Here, it is assumed that a new or upgraded drainage and sanitation system inevitably be adapted to the local infrastructure, and that change of physical conditions is practically impossible. Policies define the nature of socially agreed technical solutions. Policies are of course changeable, but they nevertheless can be considered as screening factors. If they are supportive to the type of solutions desired by the (majority of) stakeholders, these will not consider other solutions than the ones promoted by the policy. If on the contrary a policy or regulation is felt as a constraint (e.g. conservative building codes) a change may be wished, but takes much more time than allowed for the realization of the drainage and sanitation intervention under study. In both situations policies are considered compelling determinants to system feasibility.

A second distinction among objectives used in technology assessment is between *technology-specific* and *site-specific* objectives. Ideally, technology-specific objectives enable a comparison of options that is independent of the situation under study. Examples of technology-specific objectives are the skills needed for construction and operation and emission reduction efficiencies of waste water treatment systems. Site-specific objectives assess factors that determine the impact of the situation (site) on the appropriateness of the technological system. Examples are: user acceptance, affordability and willingness to pay. The objectives listed in Table 6.1 are all to a high degree *technology specific*, which does not mean that location factors can be completely disregarded.

As objectives are dependent on the situation under study and stakeholders' values and interests, there is no once-and-for-all set of objectives in drainage and sanitation improvement. The presented list, however, is useful to orient the assessment of options.

6.5 Building Blocks for the Development of Sanitation Options

Drainage and sanitation improvement may consist of upgrading existing inadequate infrastructure, or the lay out of new systems. In order to overcome a business-as-usual approach, which means a thoughtless extension of conventional systems to new areas, the MCDA method encourages workshop participants to think of as many as possible feasible alternative options, starting with simple building blocks. These are modes of water supply, toilet types, transport systems (by pipe or by cartage), septic tanks, stream separation at source and waste water treatment methods. Special attention is required for reuse possibilities.

During the multi-stakeholder workshops the building blocks are introduced during a presentation by the facilitator. The total number of conceivable system options is huge, but is strongly reduced by practical feasibility. The data-base lists three household on-site, three communal on-site and 52 off-site sanitation options. The strong emphasis on off-site options is inspired by the urban-application context of the method. Typically the data-base integrates the handling of stormwater runoff and domestic unwanted water. A descriptive grouping of the options is given in Table 6.2. The basic distinction between the options is the difference on-site, off-site (column 2) and the degree of segregation of waste streams (column 4). The maximum segregation is four streams: stormwater, grey water, faecal matter and urine (groups 10 and 12); the minimum is one stream: combined transport and treatment of sewage and stormwater (group 4). The data-base incorporates six different toilet types (Fig. 6.2). Figure 6.3 gives two examples of building blocks integrated into a system or chain. The systems represented are: (a) households with flush toilets connected to a combined sewer system with enhanced storage capacity to reduce the frequency of combined sewage overflows (belonging to group 4), and (b) households with urine-diverting dehydrating toilets with grey water treatment in a waste water treatment plant (belonging to group 10). In this option the first flush of stormwater and any flow of stormwater below a certain rate is co-treated with the grey water. Urine and faecal matter are transported by cartage. Systems with stream separation at source, such the ones of group 5, 7 and 8–12, are also called *source-oriented* systems.

6.6 Selection of Sanitation Options by Screening and Comparison

The development of system options and the assessment of consequences (also named: performance) are closely connected. During the MCDA process all participants jointly generate system options, but it is in particular the role of experts, supported by the sanitation data-base, to help select systems that are feasible in the situation under study and reject unfeasible or inadequate options. This is the process of *screening*. As noted above screening primarily checks out whether proposed system options are compatible with the infrastructure, physical conditions and policies in the situation under study. This can be done by systematically confronting options with questions about the compatibility with the situation under study. A screening aid for the 12 system groups proposed in Table 6.2 is presented in Fig. 6.4 and explained below.

Screening is important as it creates awareness among the process participants about the principal reasons for choosing or rejecting certain system options and it reduces the number of options that have to be subjected to the subsequent full, often time-consuming, comparative assessment of many other relevant objectives (Table 6.1). Screening can be applied either to a full list of drainage and sanitation options, such as the one introduced in Table 6.2, or to a short list of options composed by the participants of the MCDA process for the situation under study. An advantage of the first approach is that it systematically leads to the eligible options,

Table 6.2 Overview of drainage and sanitation options

Water supply (1)	Configuration (2)	Group number (3)	Options according to segregation of unwanted –water streams (4)	Reusable products (5)	Number of options (6)
Hand-carried water	Household and communal on-site sanitation	1	Dry anaerobic toilet systems in households and in communal toilet blocks, grey water infiltrated (three streams)	Faecal sludge and biogas (communal systems)	2
		2	Urine-diverting dehydrating toilets in households and communal blocks, grey water infiltrated (four streams)	Urine, faecal matter, biogas (in communal systems)	2
Piped water supply		3	Water-saving or high-water consumption toilets + on-site treatment (three streams)	Septic-tank sludge	2
	Off-site treatment of sewage (central and community-based)	4	Flush toilets, combined stormwater and sewage (one stream)	WWTP effluent and septic tank sludge	6
		5	Urine-diverting flush toilets, combined stormwater/brown and grey water, urine collection (two streams)	WWTP effluent, septic-tank sludge and urine	6
		6	Flush toilets, separated collection of stormwater and sewage (two streams)	WWTP effluent and septic-tank sludge	12
		7	Urine-diverting flush toilets, separated collection of stormwater/brown, grey water, and urine (three streams)	WWTP effluent, septic-tank sludge and urine	12
	Off-site treatment of source-separated streams (community-based)	8	Urine-diverting dehydrating toilets, separated collection of combined storm- and grey water, urine and faecal matter (three streams)	Urine, faecal matter, WWTP effluent	2
		9	Vacuum toilets, separated collection of combined stormwater/grey water and blackwater (two streams)	WWTP effluent, faecal sludge/black water, biogas	2
		10	Urine-diverting dehydrating toilets, separated collection of stormwater, grey-water, urine and faecal matter (four streams)	Urine, faecal matter, grey water WWTP effluent	4
		11	Vacuum toilets, separated collection of stormwater, black water and grey water (three streams)	WWTP effluent, grey water septic tank sludge, faecal sludge/black water, biogas	4
		12	As group 10: with urine-diverting flush toilets (four streams)	Urine, brown water and grey water WWTP effluent	4

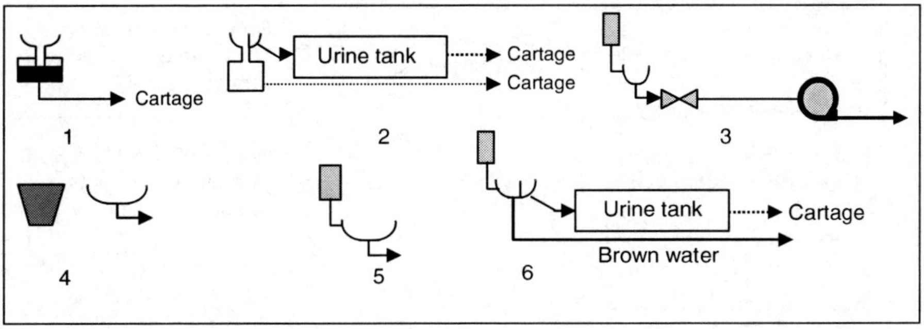

Fig. 6.2 Icons used for six toilet types. 1 – anaerobic digestion toilet; 2 – urine-diverting dehydrating toilet; 3 – vacuum toilet; 4 – pour-flush toilet; 5 – cistern-flush toilet; 6 – urine-diverting flush toilet

a. Combined sewer system with enhanced storage capacity
b. Dry urine-diverting toilet and integrated settled separated sewer system. [ST = septic tank, UT = urine tank, WWTP = waste water treatment plant].

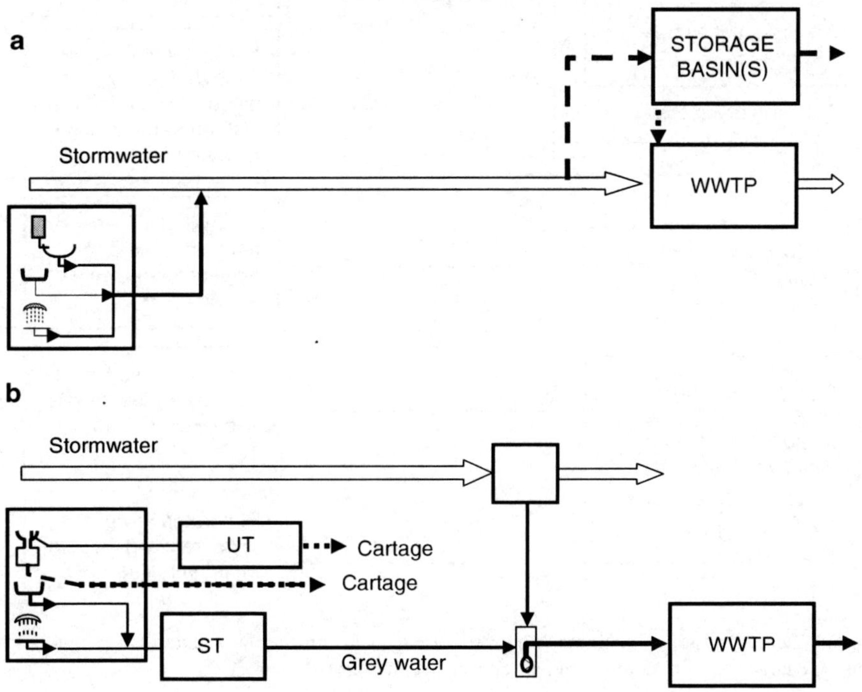

Fig. 6.3 Two examples of off-site options

eliminating many unlikely and unfeasible options from the list. It urges the participants to think of options they might not have considered otherwise. The second one however is more practical, as the participants do not have to focus their attention on

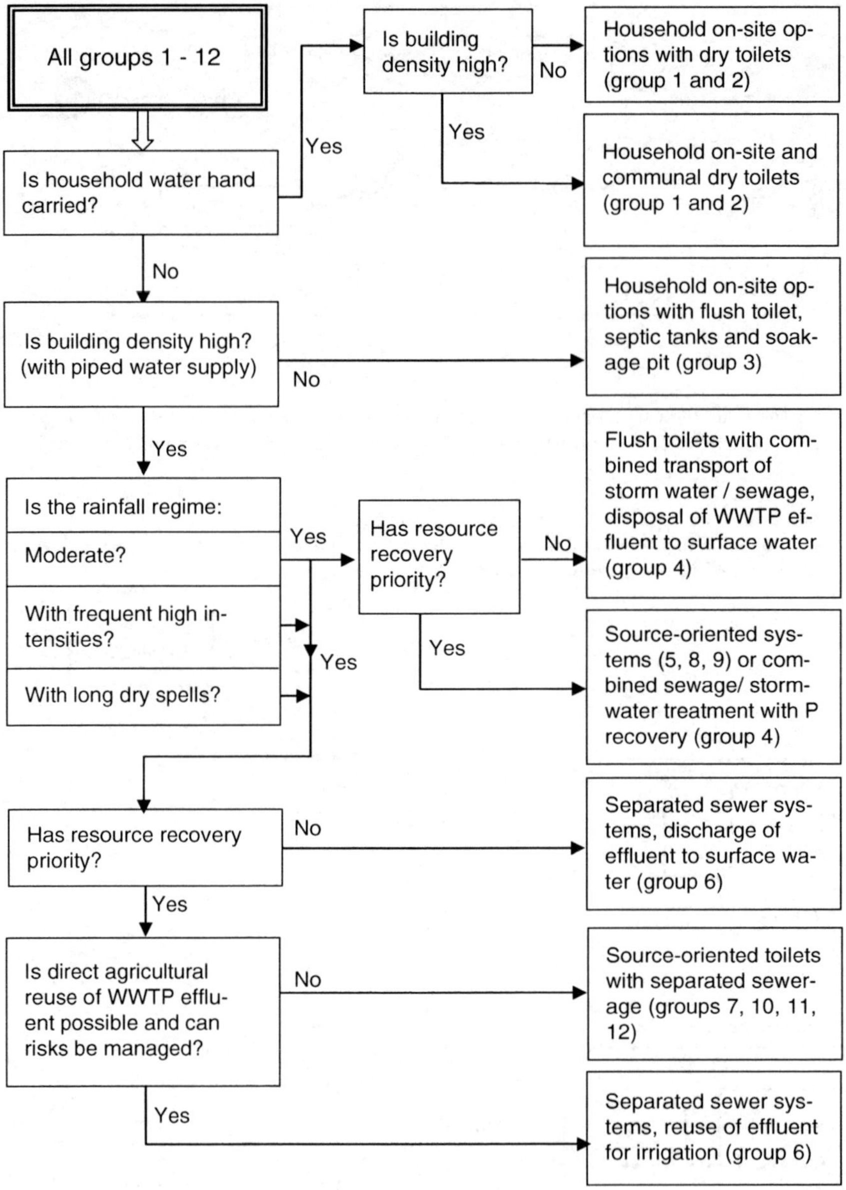

Fig. 6.4 Screening aid to distinguish between groups of drainage and sanitation options (each group is characterized by a toilet and a transport technology)

the wide range of options of which they *implicitly* know that they are unfeasible in the situation under study. Moreover, MCDA participants often come with options that are not listed, or options that are hybrids of the listed options, and for which the systematic screening aids need to be adapted.

The screening among the 12 groups (Table 6.2) is carried out by means of the following factors indicative of local conditions which determine the systems' technical functionality, that is compatibility with local infrastructure, physical conditions and compliance with the reigning policies (Table 6.1, objectives 1 and 2):

- Hand-carried water or piped water in the house
- Housing density
- Rainfall regime
- The priority given to resources saving, recovery and reuse
- The feasibility of agricultural reuse of effluent

The impact of these factors on the drainage and sanitation system choice are formulated as *questions* that single out certain system groups as feasible and others as unfeasible in the situation under study (Fig. 6.4).

The first question *'Is household water hand carried?'* distinguishes between useful application of dry on-site toilets and water-using toilets. Here, it is assumed that where households use hand-carried water only dry toilet options are feasible and flush toilet options are unfeasible. Where piped water is used dry toilets could still be used, as is the case in the North of Vietnam where households in villages with piped water have dry urine-diverting toilets and grey water is discharged to rainwater channels.[1] The question *'Is building density high?'* distinguishes between household on-site and off-site treatment. Due to a lack of space at high building density unwanted water cannot be treated on-site, except by means of communal systems, and it has to be evacuated from the built environment. Here, off-site systems have to be used.

With regard to the third question *'Is the rainfall regime......?'* the screening aid distinguishes three options: (1) *moderate*, (2) *with frequent high intensities* and (3) *with long dry spells*. This question indicates the consequences of the rainfall regime for the choice of either a combined (one pipe) or a separated (two pipes) sewage transport system. Combined sewer systems are less feasible in situations with frequent high rainfall intensities and in climates with long dry spells. The latter occurs mostly in dry climates (low annual rainfall). Here, the combined sewer system would be under-utilized during a large part of the year and there would be a high risk of clogging of gullies and pipes, due to accumulation of solids in the system. At frequent high rainfall intensities and a high annual rainfall a combined sewer system has enormous differences between the dry-weather flow and the stormwater flow. At dry weather only a small trickle of sewage finds its way through huge pipes. As long as sewage is *not* treated, but discharged directly, the only problem of combined systems is the accumulation of sewer sludge and the possibility of blockage during long periods of dry weather. This is a common situation in tropical developing countries.

---

[1] Options including dry toilets with sewered grey water belong to groups 8 and 10 of Table 6.2 (off-site treatment of grey water).

If the collected sewage *is* treated, the high flows of the mixture of stormwater and sewage occurring at high rainfall intensities have to be bypassed along the waste water treatment plant leading to a high frequency of pollution emissions (combined sewer overflows) to the receiving surface water, thus reducing the effectiveness of the treatment plant. Consequently, under conditions of long dry spells and frequent high rainfall intensities *separated* transport of stormwater and sewage is more suitable than combined transport, while under conditions of moderate and regular rainfalls *both combined and separated systems* are feasible. Under certain conditions less expensive open channels can be used for the transport of stormwater.

At any rate it is important for reasons of environmental protection and cost reduction of the drainage and sanitation system to limit as much as possible the flow and pollution of stormwater carried off by the sewer system. Infiltration and storage on-site should be facilitated and pollution with street and sewer wastes be prevented by regular cleaning of inlet points and streets. In the case of separated stormsewers much attention has to be paid to the prevention of wrongly connected domestic sewers as these can lead to considerable pollution of stormwater with sewage and thus to surface-water pollution.

Having arrived at this point of the screening a distinction on the basis of physical conditions has been made among on-site dry (group 1 and 2), on-site flush toilet options (group 3) and off-site options (4–12). Further, group 4–12 has been subdivided into the groups with combined (groups 4, 5, 8 and 9) and with separated transport of stormwater runoff and domestic waste water (groups 6, 7, 10–12).

The next main distinguishing feature within the remaining clusters of options is expressed by the question *'Has resource recovery priority?'* The answer to this question depends to a high degree on considerations of environmental policy, though also the demand of reusable products is an important factor. If such priority is not set, the waste water can be treated and discharged to surface water. Of course, even without setting a high priority, energy recovery from waste water treatment sludges by means of anaerobic sludge digestion is a standard practice at sewage-treatment plants. Where the priority of resource recovery is set, all drainage and sanitation groups both with combined and separated transport of stormwater and domestic waste water (4–12) could be organized in a way that resource recovery is realized by agricultural reuse of effluent, or by centralized or decentralized recovery of resources at source or at the waste water treatment plant (water, energy and nutrients). Here, the groups 4 and 6 (with regular flush toilets) and 5 and 7 (with urine-diverting flush toilets) have the possibility of both a centralized, large-scale, and a community-based lay out. The source-oriented options of the groups 8, 10 (dry urine-diverting toilets), 9 and 11 (vacuum toilets) and 12 (urine-diverting flush toilets) are more suitable for community-based drainage and sanitation. The latter options carry off three or four separated waste streams (water and solids) and it seems hardly financially feasible to spread such systems over cities in a centralized way.

The question *'Is agricultural reuse of waste water treatment plant effluent possible and can risks be managed?'* intends to distinguish between systems in which valorization of water and nutrients in waste water treatment plant effluent is achieved through irrigation of crops, and systems in which valuable substances are recovered in another way. This distinction is important, since agricultural reuse can be a highly effective way of effluent valorization and is far more simple than the (experimental) source-oriented systems (see also Grendelman and Huibers, Chapter 12 of this volume). Agricultural reuse presupposes a demand for irrigation water and good management of the risk of soil degradation (heavy metals, salts) and disease transmission (pathogens, persistent organic chemicals). The need for irrigation water is greatest under climatic conditions of high temperatures and low rainfall. It seems that favorable conditions for effluent reuse in agriculture coincide with conditions for separated sewer systems as defined in this screening aid (groups 6, 7, 10–12). Under conditions of a moderate rainfall regime and combined sewer systems the groups 4, 5, 8 and 9 could be applied, in which group 4 represents combined sewer systems with treatment of mixed sewage and stormwater, while the groups 5 (urine-diverting flush toilets), 8 (dry urine-diverting toilets) and 9 (vacuum toilets) are designed for the recovery of energy and nutrients by means of source-separation of waste water streams. The feasibility of source-oriented options will probably depend much on local circumstances. Group 12 for example has urine-diverting flush toilets with separated treatment of urine and faeces, and separated treatment of grey water and of stormwater. This system therefore has four source-separated streams. This group could be feasible in large buildings where the effluent of local brown water treatment is reused for toilet flushing (Werner et al. 2008).

The second stage of screening concerns the options within a group, which differ with respect to on-site pre-treatment in septic tanks and the handling of stormwater. Group 6 (separated collection of stormwater and sewage) for example consists of 12 options (Table 6.2) and distinguishes between options with and without sewage pre-treatment in septic tanks, with and without treatment of the stormwater base-flow and with and without pumping of sewage. Then, a third screening concerns the choice among several waste water treatment technologies.

As noted above screening can be carried out starting with a long general list or a short specific list of options. The aid presented in Fig. 6.4 can play a role in both approaches. When participants in the MCDA process work with a short list not all the mentioned questions may turn out relevant, but some may be useful for an inspection of the correct options selection in retrospect.

After screening the participants in the MCDA process have a list of drainage and sanitation options that are compatible with local physical and policy requirements, but whose degree of attainment of other technology-specific (Table 6.1, tertiary objectives 3–25) and site-specific objectives, such as user acceptance, needs still to be assessed and compared. For this additional assessment with regard to the technology-specific objectives the data-base provides information about the performances of the technological building blocks of the system options (toilet types and technologies for on-site treatment, sewerage and waste water treatment).

It turns out in practice that the assessment of objective attainment, or performances, evokes several questions that cannot be answered fully by means of the data available during a first stakeholder workshop. These are for example questions about emissions or costs of the selected options in the situation under study. Additional data about the consequences of the selected options will be required, which are to be collected in the follow-up of the workshop. A definitive assessment and comparison of options calls for a second workshop during which the stakeholders assess the options again, now having received more complete information about the consequences of their application in the situation under study. In Section 6.8 a recommendation for a complete MCDA process is detailed.

6.7 Evaluating Experiences with MCDA in Sanitation Improvement

Two cases have been applied to test the presupposed added value of especially the screening phase of the used MCDA method. The first case occurred within the framework of the Dutch project ISSUE-2, which works at improvement of sanitation and solid-waste management in low-income areas in 15 cities worldwide. A 2-days workshop was organized in Ho Chi Minh City, Vietnam. All invited stakeholder groups were professionals somehow involved in urban sanitation upgrading. At first the group worked at a shared vision of problems and strategies rather than on decision-making. After a general agreement on the problem analysis and the formulation of alternative solutions a strength-weakness analysis of the options was undertaken in preparation to the more quantitative appraisal of the options. Several sanitation chains were designed. To mention one example: for a certain unplanned, high density area of Ho Chi Minh City the main problem was the absence of toilets and lack of space to provide households with their own toilets. In this poor area inhabitants defecate in canals. The participants found that communal blocks with pour-flush toilets connected to (a) the existing combined sewer or (b) a separated sanitary sewer system would be the most appropriate solution. The participants quickly understood how to compose relevant sanitation chains using the icons of the building blocks created by sanitation experts (Figs. 6.2 and 6.3). Several systems and combinations of systems were proposed that were not mentioned in the data-base. Among the technology-specific objectives listed in Table 6.1 compatibility with the existing conditions; prevention of pollution, and economic efficiency were rated as most important. Typically, the importance of the objective 'resource recovery' was not ranked high, since this was rather a concern of environmental specialists than of the majority of the participating stakeholders. The most important site-specific objective appeared to be 'user acceptance'. With regard to options' assessment it was found that one workshop is insufficient to complete screening and comparative assessment of the screened options. A profound assessment of the options requires additional after-workshop study and expert input.

A second MCDA workshop experience was obtained at Makarere University in Kampala, Uganda, by a group of African PhD students working on the PROVIDE project in East Africa. In addition to the students a number of local water and sanitation experts participated. At this stage there was no participation of users from the areas under study. During this workshop sanitation options were selected and assessed for two different high-density low-income residential areas within Kampala City: Katanga and Makarere-3. The emphasis in the workshop was on generation and comparison of various drainage and sanitation options.

The group of *Katanga* has selected and evaluated six sanitation options. Four options had dry toilets (simple pit latrine, pit latrines with chemical treatment, pit latrine raised in order to overcome flooding problems, urine-diverting dry toilet with separated collection of urine and faecal matter) and two options had cistern-flush toilets, one of which with on-site treatment by means of septic tank and soakage pit, and one with separated collection of stormwater and sanitary sewage followed by a local waste water treatment plant. Two different assessments were made, one with regard to technical, environmental and cost objectives and one with user-related objectives. The Katanga group ended with a dilemma finding a contradiction between the technical and the user-related scores. Simple and chemical pit latrines obtained low technical and environmental scores, due to land use and problems with leachate and sludge handling, but the highest scores for the attainment of objectives related to user acceptance. In contrast to this, the flush toilet plus sewerage and off-site treatment option gained the highest technical and environmental score, but it was found less affordable, not to say unaffordable. The urine-diverting dry toilets were believed to be the least accepted by users.

The group of *Makarere-3* initially evaluated six options, namely the ventilated improved pit latrine, the raised pit latrine, the urine-diverting dry toilet, the twin alternating vault toilet, the aqua-privy and the bio-sanitation centre with separated treatment of excreta in an anaerobic digester and grey water by local infiltration. After having made a strength-weakness analysis this group selected the bio-sanitation centre and the household on-site urine-diverting dry toilets as most innovative and promising options. These two options were further compared by means of user-related objectives, after which it was concluded that on-site urine-diverting dry toilets would be the most appropriate new option. It was striking that the two groups in the workshop came to completely opposite assessments of the on-site urine-diverting dry toilets. This outcome was apparently not caused by different field conditions in the two study areas, but by contrary views of the participants: the Katanga group put much weight on the *assumed* lack of user acceptance of the urine-diverting toilets, while the Makarere-3 group recognized challenges in this domain but pointed at low cost and environmental advantages of low emissions and reuse opportunities.

6.8 Conclusions

A full evaluation of the MCDA method in the setting of a stakeholder dialogue includes the process of decision-making and its outcomes (Irvin and Stansbury 2004). As in the discussed cases the stage of decision-making and realization of outcomes has not yet been reached, our judgment is limited to the stages of problem analysis, formulation of objectives, generation of options, performance assessment and trade-off. During both workshops the problem reconnaissance showed the strength of the method to weld together the participants' views to a shared problem analysis and a set of solutions for drainage and sanitation improvement. An effective exchange of views and knowledge among participants was demonstrated about problems and fast learning about options and objectives. Particularly, the approach of inviting the participants to construct their systems making use of building blocks appeared useful and instructive.

In fact, the workshops integrate problem analysis with design and selection of solutions and provide a space where specialists may convince stakeholders about the importance of innovations such as closing material loops which stakeholders do not rank high at first instance. In this case the method focused on drainage and sanitation hardware, but it might also be used in the design and assessment of management-related options. The participation of stakeholders from a wider range of practices is likely leading to more specific options to choose from than if only provider-related experts had been offering options.

During the first workshops the emphasis was on problem analysis, objective setting, screening and preliminary assessment, in particular of the attainment of site-specific objectives. The stage of system assessment with respect to technology-specific objectives could be strengthened on the point of provision with chain-specific performance data through additional input of expertise, and possibly even in-depth studies, between a first and a second workshop session. At the follow-up workshop experts and non- experts should reunite for a joint assessment based on transparent data, and for the trade-off of strengths and weaknesses of the compared options.

In accordance with these experiences the procedure schematized in Fig. 6.5 is recommended. Broadly supported decisions about sanitation options in the urban setting can be reached if the assessment of system options can be integrated with value-determined measurements derived from MCDA stakeholder workshops as described in this chapter.

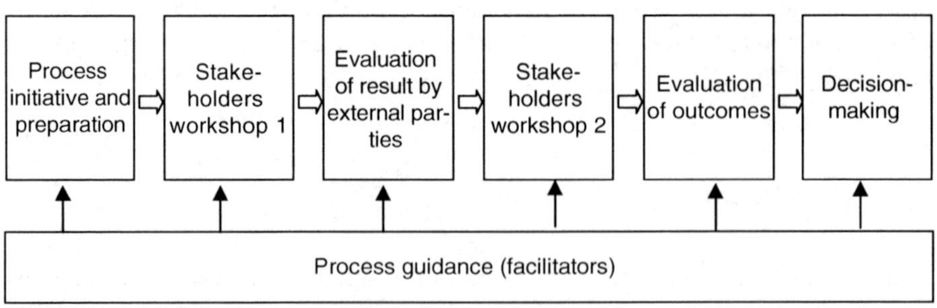

Fig. 6.5 Structure of the decision-making procedure

References

Foxon, T. J., McIlkenny, G., et al. (2002). Sustainability criteria for decision support in the UK water industry. *Journal of Environmental Planning and Management, 45*(2), 285–301.

Hammond, J. S., Keeney, R. L., et al. (1999). *Smart choices a practical guide to making better decisions*. Boston, MA: Harvard Business School Press.

Irvin, R. A. & Stansbury, J. (2004). Citizen participation in decision making: Is it worth the effort? *Public Administration Review, 64*(1), 55–65.

Lahdelma, R., Salminen, P., et al. (2000). Using multicriteria methods in environmental planning and management. *Environmental Management, 26*(6), 595–605.

Loetscher, T. (1999). *Appropriate sanitation in developing countries*. Ph.D. thesis, University of Brisbane, Brisbane, 285 pp.

PHSSDA. (2007). *Philippines Sanitation Sourcebook and Decision Aid*. Department of Environment and Natural Resources of the Philippines.

Seghezzo, L. (2004). *Anaerobic treatment of domestic wastewater in subtropical regions*. Ph.D. thesis, Wageningen University, Wageningen, 172 pp.

Van der Vleuten-Balkema, A. (2003). *Sustainable wastewater treatment*. Ph.D. thesis, Technical University Eindhoven, Eindhoven, 122 pp.

Werner, C., Olt, C., et al. (2008). Ecosan demonstration project at the head-quarters of the GTZ, Germany; Proc Sanitation Challenge, Part I-Oral Presentations p 302-3-309. *Sanitation Challenge*. Wageningen: Environmental Policy Group Wageningen University.

Zurbruegg, C., & Tilley, E. (2007). Evaluation of existing low-cost conventional as well as innovative sanitation systems and technologies. Workpackage 3 – Assessment of Sanitation Systems and Technologies. NETSSAF, Deliverable 22 & 23. Project no. 037099.

Chapter 7
Perceptions of Local Sustainability in Planning Sanitation Projects in West Africa

Jennifer McConville, Jaan-Henrik Kain and Elisabeth Kvarnström

Abstract The purpose of this study was to examine local perceptions of sustainability in the context of sanitation interventions in Burkina Faso and Mali, West Africa. Through a series of interviews with local actors criteria for sustainable sanitation were defined in the local context. These local criteria were compared with criteria found in international literature and planning practices used in two sanitation projects. The results from the interviews emphasize criteria related to behaviour change processes, while criteria in literature are either oriented toward technical assessments or project guidelines. The case studies show an attempt to merge academic and pragmatic perspectives by addressing both the technical requirements and processes of social change. As we seek to improve results within the sector it is important to start reflecting on what criteria and sustainability definitions are used in specific approaches.

7.1 Introduction

There is an increasing need for improved sanitation systems in many areas in the world, particularly in West Africa. Improvement in sanitation coverage has been targeted by the United Nations Millennium Development Goals (MDGs) because of its strong links to issues of environmental and public health, economy, and human dignity. An estimated 1.6 billion people must be able to access improved sanitation services before 2015 in order to meet the MDGs target of halving the percentage of people without access to improved sanitation (United Nations 2007). However, much of the world is not on track to meet these goals due to compounded problems of population growth, urbanization, and historically inefficient service provision. Since the start of the international development movement in the 1950s, there have

J. McConville and J.-H. Kain
Department of Architecture, Chalmers University of Technology, SE-412 96 Göteborg, Sweden
e-mail: jenmcc@chalmers.se; kain@chalmers.se

J. McConville and E. Kvarnström
Stockholm Environment Institute, Kräftriket 2B, SE-10691 Stockholm, Sweden
e-mail: elisabeth.kvarnstrom@sei.se

B. van Vliet et al. (eds.), *Social Perspectives on the Sanitation Challenge*,
DOI 10.1007/978-90-481-3721-3_7,

been numerous reports citing low success rates of water and sanitation projects (e.g. Pickford 1995; Cairncross 1992; Feachem et al. 1977). West Africa in particular is struggling to meet the demands for sanitation. This region has witnessed relative stagnation in sanitation coverage since 1990, when total access to basic sanitation was 32% (WHO and UNICEF 2006a). However, these regional figures hide significant differences between countries, for example sanitation coverage in Burkina Faso was only 13% in 2004, while in Mali it was 46% (WHO and UNICEF 2006b), although neither country is on track to meet the sanitation MDGs. The global figures can also hide significant differences in sanitation coverage between urban and rural areas within the countries. Globally if such trends continue, the world will miss the sanitation target by 600 million people (United Nations 2007).

There is a growing awareness of the sanitation problem within the international development community and initiatives such as the International Year of Sanitation (2008) mean that an increasing amount of funds will be invested in the sector over the coming years. Yet, the challenge of the MDGs is not only to achieve statistical improvements on paper (i.e. number of toilets constructed), but to do it in a sustainable manner that will lead to lasting positive change for the entire community. It is important that funds are invested in sustainable sanitation systems, since it is a very costly short-term solution to provide sanitation systems that, for different reasons, are not capable to sustain in the local reality. It is therefore important to discuss and identify critical components for achieving sustainable sanitation in design, planning and implementation processes.

In this chapter the term sanitation refers to the process of disposing of human excreta in a manner that protects public and environmental health. It is important to keep in mind that a sanitation system is more than just the user interface (the toilet), but includes the excreta collection unit; a method of transport from the site; the treatment process; and finally the end use or disposal. However, sanitation systems are also more than just physical infrastructure and therefore the definition of a sanitation system should also include the users and management frameworks that control the operation and maintenance along the entire treatment process. Since the technical sanitation infrastructure relies heavily on human interaction in both individual and communal contexts, it is closely linked to issues of culture, civil society, and economics. Therefore, a more inclusive definition has been adopted by the recently (2007) formed Sustainable Sanitation Alliance. In the context of this Alliance it is stated that sustainable sanitation systems must primarily protect and promote human health by providing a clean environment and breaking the cycle of disease and that sustainable sanitation systems must not only be economically viable, socially acceptable, and technically and institutionally appropriate, but should also protect the environment and the natural resource base (SuSanA 2007).

Despite this definition, the concept of sustainability is still very theoretical and what is actually meant by each aspect of the definition can be debated and will certainly vary depending on the context. However, the concept is useful for broadening the scope of discussion when planning and implementing sanitation systems, so that the system will be better adapted to the local context (Kvarnström and McConville 2007). In an attempt to structure the debate on sustainability, and to offer better support for the practical implementation of sanitation efforts, this chapter examines differing perspectives on what constitutes sustainable sanitation as well as different understandings of necessary requirements for achieving it. Specifically it compares the criteria and definitions used in expert planning and decision-making tools with the perspectives and experiences of local sanitation as expressed by actors in West Africa.

The objectives of this chapter are threefold: (1) to examine local perceptions of criteria for sustainability in the context of sanitation interventions in Burkina Faso and Mali, West Africa, (2) to compare those locally identified sustainability criteria with expert recommended criteria, and (3) to examine how the local and international identified criteria are used in two case studies of sanitation projects.

7.2 West African Case Studies

The following study was conducted in Mali and Burkina Faso, West Africa. The area is interesting since it has a low coverage of sanitation, while at the same time having a high prevalence of foreign aid and an advanced process of decentralization. This means there are resources available and empowerment of local government to take action for sanitation. However, given the weak results in sanitation improvement it is interesting to ask if there is a discrepancy between local and donor interpretations on the objectives of sustainable sanitation in the region. The results presented here are the first phase of a larger research project which objectives are (1) to establish if a knowledge gap exists between expert sanitation frameworks developed by academics and international development workers on the one hand, and priorities and practices implemented by local actors on the other; and (2) to explore whether such a gap may have played a role in the poor delivery of sanitation so far.

Both Mali and Burkina Faso are land-locked countries in the semi-arid savanna of West Africa. Approximately 80–90% of their populations are tied to the land through agriculture and animal husbandry. Mali and Burkina Faso rank as the 168 and 173 out of 179 countries on the Human Development Index (UNDP 2008), with GDPs per capita of US$1,058 and US$1,084, respectively. While Burkina Faso has a nominally higher GDP, it ranks lower due to other development measurements such as a lower life expectance at birth, 51.7 versus 53.7 in Mali; and school enrollment ratios, 30% in Burkina and 44% in Mali. In Mali, 40% of the inhabitants lack access to improved drinking water sources and in Burkina it is 28% (UNDP 2008). The high level of poverty, accentuated by variable climate conditions, and relatively stable political situations of these countries makes them prime candidates for foreign aid.

7.3 Methodology

The main approach of the study was to explore local perspectives relating to sustainable sanitation through an interview study with local actors. It consisted of interviews with 20 key informants within local sanitation institutions. Interviewees were asked to define 'sustainability' in the context of the local sanitation sector. Ten of the interviews were with personnel from the Regional Centre for Low-Cost Water and Sanitation (CREPA) and 10 with individuals from other institutions involved in sanitation. CREPA serves as an educational and research resource for the water and sanitation sector in French-speaking West Africa and has extensive experience working with communities to implement on-site, low-cost sanitation projects. By conducting additional interviews within this organization, the study was able to obtain perspectives from people with diverse roles and training: project managers, research specialists, technicians, and sociologists. The remaining 10 interviews covered the perspectives from other institutions involved in the sanitation sector: governmental agencies (National Office of Water and Sanitation, Burkina Faso [ONEA], General Directorate of Water Resources, Burkina Faso [DGRE], Malian Municipal Mayor; international non-governmental organizations (WaterAid, Plan, Helvetas), a local NGO (Alphalog), and international donor organizations (World Bank-WSP, UNICEF)). The interview format varied slightly depending on the situation, but in general the interviews were semi-structured, based on a list of guiding questions (Kvale 1996). Due to limited time constraints this study did not include the perspective of end-users, instead focusing on informants with decision-making power within the context of sanitation projects.

The responses from the interviews were analyzed using a "meaning categorization" approach with the aid of the qualitative research software. Meaning categorization implies that the interviews are coded into different categories in order to reduce and structure the information (Kvale 1996). The interview transcripts were coded based on categories of sustainability identified in advance from the literature reviews, and also on ad hoc categories that arose during the analysis. From these codes a number of sub-categories were discerned and eventually separated into four main categories of criteria for sustainable systems.

Interview studies have long been used in qualitative research as a structured method for capturing in-depth perspectives from a multitude of stakeholders (Kvale 1996; Miles and Huberman 1994). While the interview technique can yield deeper perspectives than a survey, it is limited by the number of people who can be included. This study is even more limited in the geographical sense that it only captures sustainability perspectives from a single sector in a specific region. However, the consistency of the results from this study and their contrast to international perspectives can exemplify the need for a more holistic and socially inclusive perspective on sustainable sanitation.

To meet the second research objective, it was necessary to identify common themes pertaining to sustainable sanitation in the global context. Therefore, the literature review aimed at synthesizing results through clustering criteria into key categories. It was found that clustering of criteria worked well when applied to similar sources, but there remained some differences across different types of publications. Therefore, in presenting the results it was necessary to specify the source of the literature, specifically differentiating between sustainability assessments of technologies, based on the use of tools such as LCA or multi-criteria decision analysis, and guidelines for sanitation development projects, derived from project evaluations and guidelines for practice.

The results were synthesized into lists of literature-based criteria which were compared to the responses from the interview study (Table 7.1).

Finally, two sanitation projects were studied to explore if and how these 'global' criteria are being used by local actors during sanitation projects. The first case was the strategic sanitation plan for Ouagadougou, Burkina Faso (PSAO) and the second was the planning process used by the international NGO WaterAid, to assist a peri-urban municipality in Mali to reaching its water and sanitation targets. Note that these cases are not full case studies, in the sense that they did not follow the in-depth and pre-specified procedures defined in case study methodology (Stake 1995; Yin 2003). Rather they focus on specific aspects of the sanitation projects to illustrate the practical application of some of the criteria discussed in the text. Information for the case studies was gathered from planning documents, project reports, and interviews with key informants.

Table 7.1 Comparison of sustainability criteria from interview study and literature review

Technology assessments	Project guidelines	Interview study
Socio-cultural		
Perception of system (importance, compatibility, familiarity)	Capacity development (institutional and stakeholders)	Capacity Development
Institutional requirements (policy, organizational structure)	Institutional network for political support/ communication	Institutional Communication
Current legal acceptability	Institutional incentives (fines, awards, enforcement)	Laws and Policy
Acceptability in current local cultural context	Adapted to demand for sanitation (convenience, cleanliness)	Cultural Acceptability
Ability to address awareness and information needs	Investment in hygiene promotion and demand creation	Awareness-Raising
Convenience (comfort, smell, attractiveness, time)	Focus on stopping open defecation (behaviour change)	Behaviour Change
Economics		
Affordability (annual and capital costs, O&M)	Affordable for the poor (willingness to pay)	Affordable
Willingness/capacity of users to pay	Apply commercial principles (enterprises, service contract)	Marketing
Local development (local resources for O&M, reusable parts)	Financial support from government	Financial Management
Cost effectiveness (cost recovery)	Subsidies for technical assistance, awareness promotion	

(continued)

Table 7.1 (continued)

Technology assessments	Project guidelines	Interview study
Technical		
System robustness (risk of failure, endure shock loading/abuse)	Technical suitability to community	Adaptation to community
Durability/lifetime	Operational efficiency for long-tem maintenance/ management	O&M requirements
Local competence for construction and O&M	Wider choice of technologies	
Local serviceability (accessible parts, technical expertise)	Use of needs-based criteria (health, poverty, environment)	
Ease of system monitoring		
Compatibility with existing systems		
Adaptability to user needs and environmental conditions		
Health		
Risk of infection from pathogens		
Risk of exposure to hazardous substances		
Environment		
Resource consumption (land, energy, materials, water)		
Environmental releases to water, air, soil		
Resource conservation (reuse, recycling)		
Impact on biodiversity, land fertility, natural systems		
Compliance with environmental standards		
Process		
	Participation (stakeholders, consensus-building, user choice)	Participation Planning
	M&E systems	M&E

7.4 The Local Perspective

Analysis of the interviews resulted in a list of locally grounded criteria for achieving sustainable sanitation, and also a tentative local definition of sustainable development within the context of sanitation. Here, 'local' is defined as equivalent to the geographic and/or operational levels where the interviewees are active; in this case urban and peri-urban areas in Burkina Faso and Mali. An important general observation is that, although the international community tends to use the inclusive definition of sustainability presented earlier, a more straightforward definition is often used among the local actors. It was explicitly or implicitly stated by all interviewees that sustainable sanitation systems are ones that will endure and continue to provide benefits after the initial stimulus, support, and funding (NGO project, awareness campaign, subsidy, etc.) have ended. In addition, the representative from Plan described sustainable sanitation as the situation where the "community exhibits ownership, where people put hygiene/sanitation lessons into practice, and where there is general cleanliness in the village".

The locally expressed criteria for achieving sustainable sanitation emphasize social-institutional aspects and behavior change. Indeed, analysis of the interview responses shows that 100% of the interviewees mention socio-cultural (including institutional) factors. Categorization of other popular responses also identified three other categories of criteria: economics, process and technical (Table 7.2). During analysis of the interview categories for environmental and health criteria were also used so as to match the categories commonly cited in the literature. Factors related to the environment were noted by three interviewees and health concerns only by one. It should be mentioned that the category of process-related criteria was not initially part of the coding during interview analysis, but developed along the way

Table 7.2 Sustainability criteria identified by interviewees

Category	Description	Cited (%)
Socio-cultural		100
Capacity development	Organizational/management skills	70
	O&M performance (proper usage)	
	Training for M&E	
	Technology appropriation	
Institutional communication	Communication plan between institutions	65
	Involvement of key leaders	
	Responsibility distribution	
Laws and policy	Functional Legal Framework	30
	Institutional policy and politics	
	Compatible with decentralization	
Cultural acceptability	Compatible with local priorities and needs	70
	Stigmas/perceptions of waste	
	Cultural value systems (dignity, gender roles)	
	Adapt to local context (social calendar, priorities, social strata)	
Awareness-raising for behaviour change	Knowledge exchange/education for informed choice	85
	Communication	
	Creating demand and awareness	
	Motivating change (authorities, early adopters)	
	Allowing time for behaviour change	
Economics		80
Affordable	Based on total life cycle costs	55
	Willingness to pay	
	Capacity to pay	

(continued)

Table 7.2 (continued)

Category	Description	Cited (%)
Marketing	Creating demand	50
	Showing the benefits/value of sanitation	
	Offering service packages	
	Creating markets, businesses, and jobs	
Financial management	Financing mechanisms (credit, subsidy, taxes)	60
	Cost recovery	
	Stability of financing	
	Capital and O&M costs	
	Locally available resources	
Process		80
Participation	Participatory approach	80
	Local organizations/leaders involved	
	User Choice	
	Ownership	
Planning	Feasibility/appropriate technology	40
	Life cycle perspective (especially considering O&M)	
M&E	Feedback and follow-up	30
	Flexible iterative approach	
Technical		50
Adaptation to local community	Environmental constraints	30
	Local capacity to replicate technology	
	Local resources (human/material) available for O&M	
O&M requirements	Local capacity/willingness to perform O&M	35
	Convenience/ease of maintenance	
	O&M requirements appropriate in cultural context	

due to the frequency that such criteria were mentioned. More specific details of what these categories contain are provided in the following sections.

7.4.1 Socio-cultural Criteria

The most common criteria for sustainable sanitation were those related to socio-cultural and institutional factors. Specifically, many (70%) of the interviewees mentioned the need to develop capacity for operation and maintenance (O&M) at both an institutional and household level. They stated that successful systems require training for proper usage and O&M, as well as management structures that will ensure correct usage through monitoring and evaluation (M&E). If the local capacity is insufficient to provide these services, implementation of a sanitation system needs to be accompanied by educational and training programs. Therefore, interviewees also cited the need to develop capacity at different levels (household, technicians, private service providers, local government), through sanitation programs that include training on proper usage, basic hygiene/sanitation knowledge, management and business skills, and organizational skills.

Another criterion that was emphasized was the need for effective communication methods and collaboration, especially between institutions (65% of interviewees). Several interviewees mentioned that this could be achieved by involving the "right" people (key leaders and individuals) and making sure that partner responsibilities were appropriately distributed. References were made to the power of traditional leaders for convincing people of the benefits of the project and having it taken seriously. By using local structures, the interviewees claimed that they strengthened communication channels and sustained the project message. Communication channels were frequently cited as methods for reinforcing capacity and strengthening motivation for behavior change through feedback sessions and learning networks.

Another form of institutional communication cited several times (30%) was the need for laws, policies and institutional frameworks for sanitation, especially ones which are compatible with the current decentralization processes. They said that appropriate laws and pro-sanitation policies are needed to push for more sanitation funding and to reinforce proper usage of facilities. It is vital that information about laws and policies is available and that capacity exists at the appropriate administrative levels to implement and enforce them. Therefore, they linked these institutional criteria to capacity development and communication/collaboration efforts between organizations. Policy and legal criteria were most often cited by interviewees representing government and donor organizations.

Finally, socio-cultural criteria also include the potentially competing factors of cultural acceptability of the systems (mentioned by 70%) and awareness-raising for behavior change (85%). Interviewees stated that proposed sanitation systems must be culturally appropriate and compatible with local priorities, especially regarding perceptions of excreta, but also respecting other social value systems. However, it was also recognized that improving sanitation practices requires behavior change and the changing of certain cultural aspects. Therefore, interviewees use awareness-raising programs to motivate behavior change and to create demand for sanitation systems through education, capacity development, and empowering community action for change. However, this must still be done within the local culture. As one interviewee put it, "it is important to understand the local culture and reinforce the good side, as well as recognize that there are cultural aspects that should be changed". Many stakeholders also noted that achieving the necessary behavior change is something that takes time; it is not something that happens overnight. In this sense, these criteria are also closely linked with the process criteria discussed below.

7.4.2 Economic Criteria

Economic criteria identified during the interviews focused on issues of affordability (55% of interviewees), marketing (50%), and financial management (60%). Due to generally high levels of poverty in the region, most actors agree that the cost of the system needs to be estimated by life cycle costs which include O&M, and be matched with what people are willing and able to pay. Essentially the cost of the systems offered must be low enough that people are willing to buy it, and/or pay for its services. One thought on designing sanitation systems was to let the market dictate the technologies offered (and hence the sustainability of systems), based on the assumption that users will chose that option that they could afford to operate and maintain.

However, most interviewees agreed that selling sanitation systems to people is also highly dependent on how they prioritize sanitation and how they choose to spend their money. They recognized that user choice could be influenced by awareness-raising efforts and social marketing efforts to get people to prioritize sanitation. Interviewees pointed to the role of market forces in sustainability, either through advertising or by improving the market for sanitation ventures. Marketing can create a desire for sanitation by advertising the benefits of sanitation or purely popularizing the idea. Several actors pointed to the need to improve the sanitation market by developing the supply chain and local capacity to provide services (construction and skilled labor, business management, organizational structures) and linking sanitation to other markets (and advertising channels) through service packages connected to other technologies.

Financial management was among the most frequently mentioned economic criteria, underpinning the issues of affordability and marketing. It was generally agreed that financing mechanisms for sanitation programs are needed and that they must cover both installation and O&M costs. However, there was disagreement on which methods to use for guaranteeing funding of service provision and assisting households with installation costs (e.g. taxes, subsidies, or credit systems). Still, the need for cost-recovery and continued financing of the system from installation through long-term O&M was often stressed. Implicated in this discussion is the idea that sanitation is a public service and that the financial management is linked to public programs and funding sources. However, the full debate of public or private management did not enter into the interview discussions, and the agreement around this issue centered on the need to plan for the long-term financial viability of the system.

7.4.3 Process Criteria

These criteria relate to the process of implementing sanitation systems or programs, and include the use of participatory approaches, planning, and M&E mechanisms. A majority of interviewees (80%) cited participatory methods as necessary for sustainability. Community participation is championed as a way to develop ownership, community empowerment, and get people to buy into sanitation options that will meet their needs. Some interviewees (40%) linked participation to planning processes that were designed to include different groups of stakeholders so as to identify diverse sets of priorities and drivers for sanitation improvements. An appropriate planning process, according to them, will assure that the outcomes of the project will meet the needs of the people, but also that it can be implemented and operated correctly in the local context. Interviewees described the need for participation and planning processes that are flexible, iterative, and include a range of stakeholders, a variety of technical choices, and planning from project start through final O&M. Others (30% of interviews) linked the process of participation to that of soliciting and responding to feedback through monitoring and evaluation (M&E) that will keep the system on track and continually identify and correct problems throughout the life of the system.

It is important to note that process criteria are not independent of the other criteria; rather they are closely linked to methods for communication, capacity development, empowerment, understanding of cultural issues, and discussions leading to informed user choice. However, due to the frequency with which these issues were cited in the interviews they were considered important enough to be mentioned as a particular set of criteria.

7.4.4 Technical Criteria

Technical criteria were not mentioned as often as the former categories (only by 50% of interviewees), and when they were it was often in relation to socio-cultural issues, such as adaptation to the local culture and environment, ease of repeatability, convenience, and local capacity and/or willingness to perform O&M. It was recognized that the choice of technical system will affect sustainability, but the process of making the choice and the participatory dialogue surrounding the choice were cited more often than specific technical indicators of success.

7.5 Expert Perspectives in Literature

As previously stated, there is a wide body of literature that attempts to categorize and generalize criteria necessary for a sanitation system to be sustainable (Box 7.1).

There are generally two perspectives regarding criteria for sustainable sanitation in literature: technology assessments and project guidelines. Technology assessment criteria are used to assess the impacts of the technologies themselves or make comparisons between them, using tools such as life cycle assessment (LCA) and multi-criteria decision analysis (MCDA). Criteria used in these assessments and commonly cited among academics in international journals and conference papers fall into five main categories: health, environment, technical, economic, and socio-cultural factors.

The second main source of criteria for sustainable sanitation comes from the international development community, presenting lessons learnt and recommendations for scaling-up sanitation projects. Information sources for this type of literature come from organizations such as the World Bank Water and Sanitation

Box 7.1 Criteria for sustainable sanitation

Acknowledgment of the need for sustainable sanitation and improvements in the current approaches has led to numerous recommendations and frameworks for improving the success of sanitation interventions. Although it is recognized that sustainability is highly context dependent and site specific, there is a wide body of literature that attempts to categorize and generalize criteria necessary for a sanitation system to be sustainable (Mukuluke and Ngirane-Katashaya 2006; Bracken et al. 2005; Balkema et al. 2002; Dunmade 2002; Hellström et al. 2000). Criteria for health, technical, economic, environmental, and socio-cultural aspects have been incorporated into sustainability assessments and decision-support models using tools such as life cycle assessment (LCA) and cost-benefit analysis (McConville and Mihelcic 2007; Van der Vleuten-Balkema 2003; Lundin and Morrison 2002). Increasingly, criteria are also being tied to process-oriented approaches in planning and implementation, such as Open Planning of Sanitation Systems (Kvarnström and af Petersens 2004) and Household-Centered Environmental Sanitation (Eawag 2005).

Basic categories of criteria:

Economic	Technical	Socio-cultural
Environment	Health	

Program (WSP), the United Nations Development Program (UNDP), Canadian International Development Agency (CIDA), Engineers Without Borders and United States Peace Corps. A synthesized list of criteria from project guidelines shows that this literature often neglects health and environmental criteria and instead emphases the categories of socio-cultural, economics and technical factors, especially those related to capacity and institutional development; affordability and financing mechanisms; and appropriate technology. Project guidelines also introduce the idea that criteria for sustainable sanitation need to capture the planning and implementation processes by their focus on participatory methods.

When comparing the findings from the literature with the results from the interviews (Table 7.1), one can notice the higher focus on technical, health, and environmental criteria in the technology assessment literature. In contrast, the criteria found in the development project literature more closely matches the criteria present in the interview material, with its emphasis on socio-economic, cultural and procedural issues. This is probably because the background context for the interviewees more closely resembles that of the international development community. Here, the importance of sustainability criteria becomes less related to the subject of sanitation itself, than it does to the social context.

7.6 Case Study Perspectives

One of the main applications of criteria is for planning and decision-making. Therefore, another way to understand these criteria is to see how actual projects are planned and implemented. In other words, how well do the actions taken by local actors match what they define to be necessary for sustainable sanitation? Two sanitation projects from Burkina Faso and Mali are used to explore this issue.

7.6.1 WaterAid, Mali

WaterAid is an international NGO that has been working in Mali since 2001 with the mission to provide the poor with access to water, sanitation and hygiene promotion. They work specifically at the municipal level to enable local leaders to develop and implement sector plans for meeting water and sanitation needs. The case of Kalabancoro, Mali is typical of the process used by WaterAid in the region (Box 7.2).

Box 7.2 WaterAid planning approach

The approach of WaterAid is to provide support for local governments through the process of decentralization and good governance that will enable them to meet the needs of the water/sanitation sector and the MDGs. They use participatory methods to build social capital by reinforcing the capacities of the different actors, especially regarding decision-making capabilities and the management of local affairs.

The development of the sector plan consisted of five steps:

- Preparatory activities
- Data collection
- Evaluation of data and creation of thematic maps
- Formulation of a plan
- Validation of the plan with the local population and action planning

The preparatory activities include initial meetings to bring stakeholders into the process. The idea is to build consensus on expected outcomes, and to train local field workers in participatory data collection techniques. This initial step allows WaterAid to identify the principle actors in the water/sanitation sector and their capacities (e.g. education levels of municipal council, organizational structure of water associations, and generally education level of the population). This is followed by the collection of data related to local water and sanitation conditions, physical surveys of existing infrastructure and environmental conditions, and socio-economic perspectives of demand and need for sanitation. The data are collected by locally trained hygienists and technicians who interact directly with

(continued)

Box 7.2 (continued)

the community. The collected data are initially processed by experts within WaterAid who use cartography and GPS systems to display the collected data. It is then validated by the community during two stakeholder workshops.

These workshops lead into the planning process by generating discussions on priority actions, feasibility and acceptability of options, and their potential impacts. During the planning process, WaterAid works with a planning committee, headed by the mayor and a representative of the interests of the community to provide training on organizational and planning techniques. The final document details a yearly plan of activities and budgets, including descriptions of improved latrines to be provided to households, awareness-raising and sanitation promotion programs, training and equipping masons, and the creation of production centres for latrine slabs. Financing and M&E of results remain the responsibility of the municipality. The final document is voted on by the municipal council and becomes the water/sanitation policy document for the municipality.

A comparison linking the planning steps used by WaterAid to the sustainability criteria identified in the interview study is presented in Table 7.3. The approach covers the socio-cultural criteria well, especially those for capacity development at the institutional level which are included in several steps. Cultural acceptability is also addressed through the participative evaluation process for the technical data and formulation of the plan. Working through the municipality assures that the legal and policy criteria are satisfied since validation of the plan makes it part of the municipal policy. Awareness-raising activities were included in the final planning document and project budget, even if few specific activities or targets for behavior change were written into the document.

In addition to a focus on capacity development and socio-cultural issues, the approach focuses heavily on participative processes. Participation of various stakeholders is included throughout the five step planning process. Process criteria for M&E are less well developed since WaterAid's intervention is mostly during the planning phase of the project. However, it is noted that participatory M&E should be addressed following the implementation of the plan.

The Malian case study project is found to be weak, however, in incorporating economic and technical criteria. On the technical side, the approach does not develop technical designs either through suggestions of appropriate systems or O&M management schemes (rather it encourages promotion of 'improved' sanitation). It does suggest organizing enterprises for emptying pit latrines, but the lack of other technology specifications means that O&M requirements are mostly neglected. Combined with the inadequate assessment of technical options, issues of affordability and financing appear to be given low priority. The final document suggests

Table 7.3 Specifics of steps used by WaterAid and ONEA in sanitation projects

Criteria from Interviews	WaterAid	ONEA
Socio-cultural		
Capacity building	Preparatory activities Formulation of plan Validation of plan	Pilot phase
Institutional collaboration	Preparatory activities Information collection Formulation of plan	Preliminary accords Situational analysis Pilot phase
Laws and policy	Validation of plan	Official adoption of plan
Cultural acceptability	Evaluation of data Formulation of plan	Situational analysis Pilot phase
Awareness-raising	To be address during implementation (promotional activities)	Pilot phase Stakeholder workshops
Economics		
Affordable		Situational analysis
Marketing	To be address during implementation (formation of enterprises)	Pilot phase
Financial management		Situational analysis Pilot phase
Process		
Participation	Preparatory activities Information collection Evaluation of data Formulation of plan Validation of plan	Preliminary accords Situational analysis Pilot phase Stakeholder workshops Strategic plan
Planning	Preparatory activities Formulation of plan Validation of plan	Stakeholder workshops Strategic plan Final plan
Monitoring and evaluation	Following implementation of plan	Pilot phase Stakeholder workshops
Technical		
Adaptation to local community	Information collection Evaluation of data	Situational analysis Pilot phase Detailing strategic plan
O&M requirements		

marketing measures to be done during implementation, but otherwise economic questions of affordability and financial management are poorly addressed. The data collection step included some socio-economic information, but not specifics on willingness or capacity of users to pay for sanitation improvements. Although the planning document includes a budget for the activities, it does not specify how the projects will be financed. It is generally assumed that the municipality will find their own means of financing the projects.

7.6.2 National Office of Water and Sanitation, Burkina Faso

The growing problems of sanitation in the city of Ouagadougou pressurized the government of Burkina Faso to draft a Strategic Plan for Sanitation (PSAO), detailed in Box 7.3. Implemented in urban areas by the National Office of Water and Sanitation (ONEA), the PSAO rests on a strategic approach to devise sanitation solutions that are demand-responsive, flexible, involving the active participation of stakeholders, and an innovative use of a sanitation surtax on the drinking water for financing the program (WSP 2002; ONEA 1993).

The relationship of the ONEA planning process to the criteria identified in this study is also shown in Table 7.3. Similar to WaterAid, ONEA has a strong focus on process criteria and in particular the participatory approach. In contrast to WaterAid however, the ONEA approach covers more of the criteria and specifically some that WaterAid missed. For example, one of the strengths of the ONEA approach lies in financial management and the focus on on-site sanitation as an affordable option for the population. The success of this approach is linked to the establishment of institutional arrangements between the government and ONEA that allow for financial (control of sanitation surtax) and managerial independence over the program direction. This arrangement makes the technologies offered more affordable through subsidies and meets the criteria for a financial management scheme for continuing the life of the program. In addition, key features of the approach include the use of social development tools to promote education and demand for appropriate sanitation solutions. By offering households a variety of options and educating them on their choices, the project increases chances for meeting the goals of cultural acceptability, awareness-raising and stakeholder participation.

There are some areas however where the ONEA approach does not match the criteria identified in the interview study. Although the planning documents for the PSAO contain more technical designs and cost estimates than the WaterAid case, there is a lack of strategy for O&M, so that faecal sludge management and the emptying of latrines remain a problem (Koanda 2008). In addition, the large cost difference between latrine options means that households with limited income essentially have only one option that is affordable; for example a standard VIP latrine (US$100) costs two to three times more than rehabilitating a traditional latrine (WSP 2002). There is also evidence that benefits and subsidies go mostly to the middle class who have more access to information (ibid.), which indicates weaknesses in communication, participatory frameworks and inequity in capacity development efforts. According to the ONEA representative there are also economic challenges to overcome; the sanitation tax is limited and costs of technologies and promotional activities are high, making it difficult to reach all sectors of the city. Since the program is run through a national level agency there has been limited organization/institutional capacity development at the municipal level and there continues to be discrepancies between the responsibilities that are given to the city of Ouagadougou through the process of decentralization and the flow of financial resources (WSP 2002).

Box 7.3 ONEA approach

The planning approach used for the PSAO and subsequently recommended by ONEA for other urban areas in Burkina Faso is composed of a series of steps and baseline studies (ONEA 2007). The first step lays out the *preliminary agreements* (1) between the municipality and ONEA. During this step a project team, containing both technical and sociological experts, is established for guiding the process. The cornerstone of the strategic approach is the *situational analysis* (2) that will allow the plan to be adapted to the local constraints and opportunities. The analysis starts with an initial workshop to build cohesion with principal stakeholders (municipality, ONEA, central administration, private sector, NGOs, current projects, development partners) and identify existing sources of information. This is followed by a series of baseline studies to obtain specific information on existing types of sanitation, user willingness to pay, living conditions, geography/soils, technology strategies, studies of desires/priorities, environmental feasibility, institutional and financial situation. The execution strategy is then defined during a *pilot phase* (3) where possible sanitation options identified by ONEA technicians are demonstrated in specific areas of the city. In addition to assessing technical feasibility, the pilot project is a method for clarifying the roles and responsibilities of different institutions implicated in the process, and judging the need for capacity reinforcement (human resources, organizational structures). This stage also includes social marketing of sanitation and an elaboration of tools that will be used for project implementation. For the PSAO, a demonstration and promotional project for on-site sanitation was piloted in two neighborhoods over 24 months starting in 1992. During this time a series of awareness-raising events were conducted to inform the community about the proposed solutions and the conditions of financing; including group meetings, household visits, and media publicity. ONEA collaborated with local NGOs to train field workers in participatory methods for sanitation promotion and masons in proper construction techniques. The entire process also included (4) *stakeholder workshops* (minimum of three) to confirm the results of the situational analysis and demonstration project with the community. Based on the pilot project and stakeholder input, possible sanitation options are identified by ONEA technicians and details are laid out in a *strategic plan* (5). The strategic plan also includes a plan for monitoring and evaluation to assure the pertinence, efficiency and sustainability of the actions taken. The *final plan* (6) is voted on by the municipal council and becomes the political policy for sanitation in the city.

In the case of the PSAO, the final plan contained three main components:

- On-site sanitation for 80% of households
- School latrines and hygiene education
- Sewerage network for the city centre and industrial area

7.7 Conclusions

Overall, this study highlights some key differences in perspectives between local practitioners in the field of sanitation provision and those expressed in criteria-based sanitation assessment tools developed by external experts, especially as regards the importance of social versus technical criteria. To some extent, this divergence stems from fundamental differences in the base definition of sustainable sanitation. Local actors in the interviews focused on the concept of sanitation as a continuously functioning system that can be managed without outside support. Compared to the often technology-oriented and theoretically holistic definitions of expert literature, actors in the West African sanitation sector stress the need to focus on social issues such as reinforcing behavior change, developing local capacities and establish long-term financing mechanisms. For them, achieving sustainable sanitation systems is a process rather than choice of technology. The importance of these process-related criteria can be seen in how closely they are linked to other key issues for sustainability, such as methods for communication, capacity development and empowerment, understanding of cultural issues, and discussions leading to informed user choice.

These different definitions are also seen in different prioritizing of sustainability criteria, for example those related to health and environmental factors. In contrast to expert literature, these two factors were rarely mentioned related to sustainability in the interviews. Expert literature often highlights these aspects since they are considered the primary purpose of sanitation and key for the technical system design. The local perspective also recognizes the importance of health and environment, but perhaps they do not mention them since it is assumed that a functioning, sustainable system will provide these benefits, and that by fulfilling the social, economic and technical criteria, health and environmental improvements will automatically be achieved. From another perspective, the answers collected through the interviews tend to agree with criteria from development project guidelines, perhaps because, at a local level, these actors are more closely linked to the implementation process in the field. This is not to say that any perspective is "correct", but rather that there are many ways of looking at the same problem. Open dialogue is needed between the differing perspectives in order to understand the whole picture.

The sanitation projects studied here illustrate the concrete application of many of the criteria identified in this study by the interviewed practitioners in the field. Both WaterAid and ONEA mix participatory planning techniques with technical baseline studies to address sustainability issues in social, economic and technical terms. However, neither project applies all of the criteria from neither the literature review nor the interview study. WaterAid focuses on capacity development but makes fewer provisions for technical and economic issues. In contrast, technical and financing measures are more clearly defined by ONEA, although they still struggle with O&M issues and extending the program to reach all stakeholders. However, it is still too early to tell what the real sustainability of these programs will be. The WaterAid project has yet to be fully implemented, and there has been no comprehensive evaluation of the long-term results from the ONEA program. It is possible that sustainability can be achieved without addressing all of the criteria that have been put forth from either local or global perspectives. However, as we seek to improve results within the sector it is important to start reflecting on what criteria and sustainability definitions are used in specific approaches; what worked, and what can be improved.

The study shows that there is a vast amount of experience, expertise, and differing perspectives among the various actors in the field of sanitation. Future research should expand this type of questioning to include perspectives from a broader range of local and global stakeholders, and compare them to additional projects and planning frameworks. For example, this study focused on professionals in the sanitation sector, but it would be both important and interesting to capture user perspectives in future studies.

The solution to the sanitation challenge will most possibly involve understanding perspectives from the wide range of involved stakeholders, academics and practitioners, and merging them into complementary and more comprehensive approaches. However, even more necessary is to start understanding how approaches to the planning of sanitation can be better grounded in local realities. This means shifting from a one-approach-fits-all mindset to locally grounded criteria that lead to informed choice between the numerous, potentially valid, approaches to planning and sanitation systems design.

Acknowledgements The authors would like to thank all the participants in the interview study for their time and insights, as well as the staff at CREPA headquarters for logistical support and hospitality. Financial support for Ms. McConville was provided by the US National Science Foundation Graduate Research Fellowship Program.

References

Balkema, A. J., Preisig, H. A., Otterpohl, R., & Lambert, F. J. D. (2002). Indicators for the sustainability of wastewater treatment systems. *Urban Water, 4*, 153–161.

Bracken, P., Kvarnström, E., Ysunza, A., Kärrman, E., Finnson, A., & Saywell, D. (2005). Making sustainable choices – Development and use of sustainability oriented criteria in sanitary decision-making. *Proceedings of the 3rd international ecological sanitation conference* (pp. 486–494).

Cairncross, S. (1992). *Sanitation and water supply: Practical lessons from the decade. UNDP-World Bank water and sanitation program discussion paper no. 9*. Washington, DC: The World Bank.

Dunmade, I. (2002). Indicators of sustainability: Assessing the suitability of a foreign technology for a developing country. *Technology in Society, 24*, 461–471.

Eawag: Swiss Federal Institute of Aquatic Science and Technology. (2005). *Household-centred environmental sanitation: Implementing the Bellagio principles in urban environmental sanitation*. Duebendorf, Switzerland: Eawag.

Feachem, R., McGarry, M., & Mara, D. (1977). *Water, wastes, and health in hot climates*. London: Wiley.

Hellström, D., Jeppsson, U., & Kärrman, E. (2000). A framework for systems analysis of sustainable urban water management. *Environmental Impact Assessment Review, 20*, 311–321.

Koanda, H. (2008). Strategic sanitation planning approach in Burkina Faso: Lessons learnt after 10 years of implementation and promising improvements. *Abstract Volume, World Water Week in Stockholm,* August 17–23, 2008.

Kvale, S. (1996). *An introduction to qualitative research interviewing*. Thousand Oaks, CA: Sage.

Kvarnström, E. & af Petersens, E. (2004). *Open planning of sanitation systems, report 2004–3*. Stockholm, Sweden: EcoSanRes Program, Stockholm Environment Institute.

Kvarnström, E., & McConville, J. (2007). Sanitation planning – a tool to achieve sustainable sanitation? In P. Wilderer et al. (Eds.), *Proceedings of water supply and sanitation for all*. Conference held in Berching, Germany, September 26–28, 2007.

Lundin, M. & Morrison, G. M. (2002). A life cycle assessment based procedure for development of environmental sustainability indicators for urban water systems. *Urban Water, 4*, 145–152.

McConville, J. R. & Mihelcic, J. R. (2007). Adapting life cycle thinking tools to evaluate project sustainability in international water and sanitation development work. *Environmental Engineering Science, 24*(7), 963–974.

Miles, M. B. & Huberman, A. M. (1994). *Qualitative data analysis: An expanded sourcebook*. Thousand Oaks, CA: Sage.

Mukuluke, J. Z. & Ngirane-Katashaya, G. (2006). Exploring sustainability of sanitation systems: Social-cultural acceptability analysis of technology options for Kampala's peri-urban areas using multi-criteria decision analysis. *Journal of Engineering and Applied Sciences, 1*(4), 445–455.

Office National de l'Eau et de l'Assainissement (ONEA). (2007). *Note d'orientation pour l'élaboration de plans stratégiques d'assainissement des eaux usées et excréta dans les centres urbains et semi urbains du Burkina Faso*. Burkina Faso: National Office of Water and Sanitation.

Office Nationale de l'Eau et de l'Assainissement (ONEA). (1993). *Plan Stratégique d'Assainissement des eaux usées de la ville de Ouagadougou (PSAO)*. Burkina Faso: National Office of Water and Sanitation.

Pickford, J. (1995). *Low-cost sanitation, a survey of practical experience*. London: Intermediate Technology Publications.

Stake, R. E. (1995). *The art of case study research*. Thousand Oaks, CA: Sage.

Sustainable Sanitation Alliance (SuSanA). (2007). *Short vision document*. Retrieved March 31, 2009, from http://www.susana.org/index.php/lang-en/vision/42-vision/53-what-is-sustainable-sanitation

United Nations. (2007). *The millennium development goals report*. New York: United Nations. Retrieved May 6, 2008, from http://www.un.org/millenniumgoals/pdf/mdg2007.pdf

United Nations Development Programme (UNDP). (2008). *Human development indices: A statistical update 2008 – HDI rankings*. Retrieved March 31, 2009, from http://hdr.undp.org/en/statistics/

Van der Vleuten-Balkema, A. (2003). *Sustainable wastewater treatment – developing a methodology and selecting promising systems*. Doctoral thesis from Technische Universiteit Eindhoven, Eindhoven University Press, The Netherlands.

Water and Sanitation Program-Africa Region. (2002). *Field note: The Ouagadougou strategic sanitation plan – an holistic approach to a city's problems*. Nairobi, Kenya: WSP Africa Region.

WHO and UNICEF. (2006a). *Meeting the MDG drinking water and sanitation target: The urban and rural challenge of the decade*. Geneva, Switzerland: WHO Press.

WHO and UNICEF. (2006b). *Joint monitoring program for water supply & sanitation*. Retrieved May 9, 2008, from http://www.wssinfo.org/en/welcome.html

Yin, R. K. (2003). *Case study research design and methods* (3rd ed.). Thousand Oaks, CA: Sage.

Chapter 8
Interactions Between Urban Forms and Source-Separating Sanitation Technologies

Franziska Meinzinger, Volker Ziedorn and Irene Peters

Abstract Up to now there is little experience regarding the retrofitting of source-separating sanitation technologies into existing housing stocks. This paper examines how specific characteristics of urban form can impact on the implementation of new sanitation technologies. A typology for residential areas in Germany was developed and exemplified in nine reference neighbourhoods in the city of Hamburg. The typology is based on physical characteristics like floor-space index, plot size and housing density. Four different source-separating sanitation systems were defined including urine separation and separate collection of black water in vacuum toilets. The suitability assessment of the four systems regarding their implementation into the different types of residential areas revealed that population density, housing layout and housing types are crucial factors. Also, possible impacts of socio-economic factors like family distribution or income are considered. As an example how urban form could affect investment costs, a cost assessment of the implementation of urine separation into the reference neighbourhoods was carried out. Housing density, population density and type of dwelling are factors influencing the specific investment costs. A reduced housing density results in increased demand for either pipes or storage containers making the implementation in less densely populated areas more costly.

F. Meinzinger (✉)
Institute of Wastewater Management and Water Protection, Hamburg University of Technology, Eissendorfer Str. 42, 21071 Hamburg, Germany
e-mail: f.meinzinger@tuhh.de

V. Ziedorn
egeb – Business Development Company Brunsbüttel, Elbehafen, 25541 Brunsbüttel, Germany
e-mail: ziedorn@egeb.de

I. Peters
Urban Technical Infrastructure Systems, Department of Urban Planning, HafenCity University, Schwarzenbergstr. 95, 21073 Hamburg, Germany
e-mail: irene.peters@hcu-hamburg.de

B. van Vliet et al. (eds.), *Social Perspectives on the Sanitation Challenge*,
DOI 10.1007/978-90-481-3721-3_8,

8.1 Introduction

Currently, we are seeing a paradigm shift in waste water management. The constraints of conventional systems that are based on centralized structures and linear flows become evident. Conventional wastewater systems show a relatively high degree of inflexibility, making it difficult to respond to changing conditions like demographic transformations. The use of large amounts of water for toilet flushing is considered a waste of precious resources. Last but not least, the mixing of wastewater flows with different characteristics impedes selective treatment and use of wastewater constituents.

Researchers and practitioners all over the world look for alternatives towards more integrated and resource-efficient concepts. Water recycling and the recovery of nutrients and energy are main objectives of these new concepts. In order to achieve this, source-separating technologies and separate treatment of different wastewater flows like urine, faeces and grey water are advocated (Otterpohl 2001). At present, in the European context these new technologies are generally introduced in new buildings only. Pilot projects are implemented in countries like Sweden, Switzerland, Germany and the Netherlands, promoting new technologies like the separation of urine and its subsequent use as plant fertiliser or the digestion of black water for energy production. Since these new technologies are normally implemented in new housing developments, the specifications related to source-separating sanitation technologies can be considered already during the design phase.

However, there is little experience regarding retrofitting housing stocks with source-separation concepts (Schäfer and Rudolph 2001). Yet, focusing on existing housing areas becomes more and more important. In Germany, more than 60% of the construction volume is currently directed towards refurbishment of existing buildings. In 1980, this ratio has been 32% (Zink 2007). In addition, treatment units become smaller and more advanced and specific, so that more decentralized solutions become feasible for large scale applications.

The objective of this study is to investigate the impact of different types of urban form on the implementation of new sanitation concepts based on a theoretical approach. Important characteristics of different residential areas are defined to assess the suitability of new approaches to waste water management and to identify possible constraints. The aim is an assessment of the interaction of different types of urban form and source-separating systems rather than a rating of the latter ones. In the following sections a typology of urban form is illustrated. Reference neighbourhoods for the nine selected types of urban form are identified in order to have a more in-depth analysis. This includes a view on the possible impacts of socio-economic factors. In conclusion, the general suitability of four source-separating systems with regard to the type of urban form is checked. In addition, the interactions of urban form and sanitation systems are exemplified by a cost comparison of the implementation of urine separation. For the cost analysis two alternative options are studied. One is based on a fixed collection cycle and the other one is based on a flexible collection cycle dependent on the filling rate of the urine storage containers.

8.2 Source-Separating Technologies and Systems

Several innovative technologies exist or are currently under development aiming at more resource-efficient waste water management. For example, water recycling units are available that treat the grey water (i.e. water from kitchen and bathrooms) for reuse in different household applications like toilet flushing or washing machines. Other technologies aim at the separation of nutrient-rich flows, which can be used as plant fertilisers. Also flows that have a high content of organic matter like faeces or brown water (i.e. faeces together with flush water) are a valuable waste water stream suitable for resource-recovery. Although source-separating sanitation concepts focus on separate flows, the interplay of different wastewater flows is important to consider. Therefore, the focus should be not only on single flows but on systems as a whole. For the purpose of this study, different waste water management systems are defined, covering a variety of different source-separating technologies. The flows under consideration include urine, faeces and greywater from households. In addition, rain water is also an important flow for the conceptualization of an integrated water/waste water system. However, this study does not deal with rain water flows, since decentralized rain water management including rain water infiltration or use is already a widely researched field and many implementations and experiences are available. Moreover, the interrelation of urban form and rainwater management has been studied elsewhere (e.g., Herzer 2004; Löber 2001). In total, four systems representing different approaches to resource-efficient sanitation systems are defined and used for this study. The systems range from a modification for only one single flow (i.e., separation of urine) to a totally self-sufficient system with decentralized treatment units for all flows (see Figs. 8.1, 8.2, 8.3, 8.4

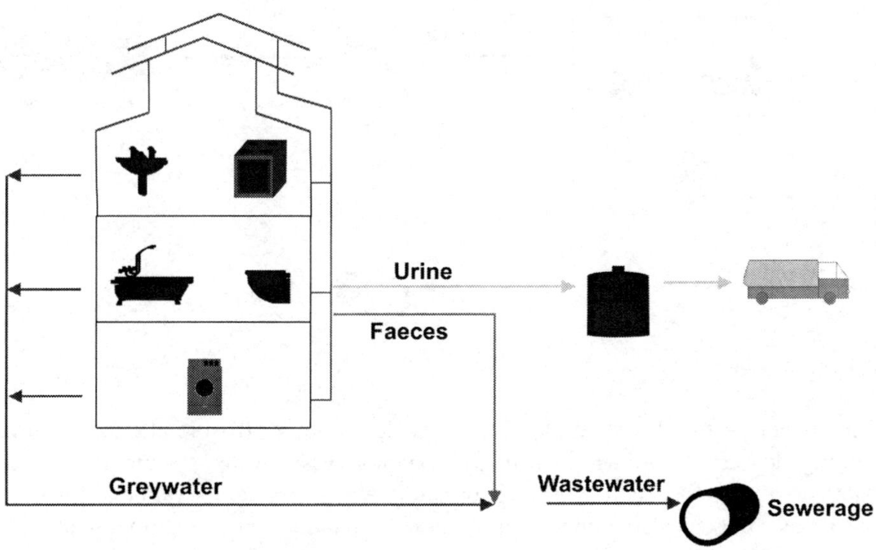

Fig. 8.1 System 1

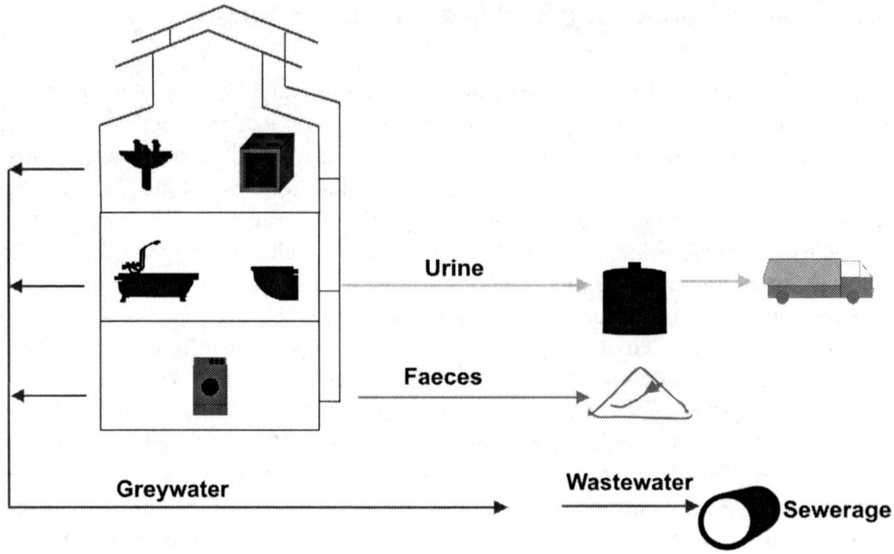

Fig. 8.2 System 2

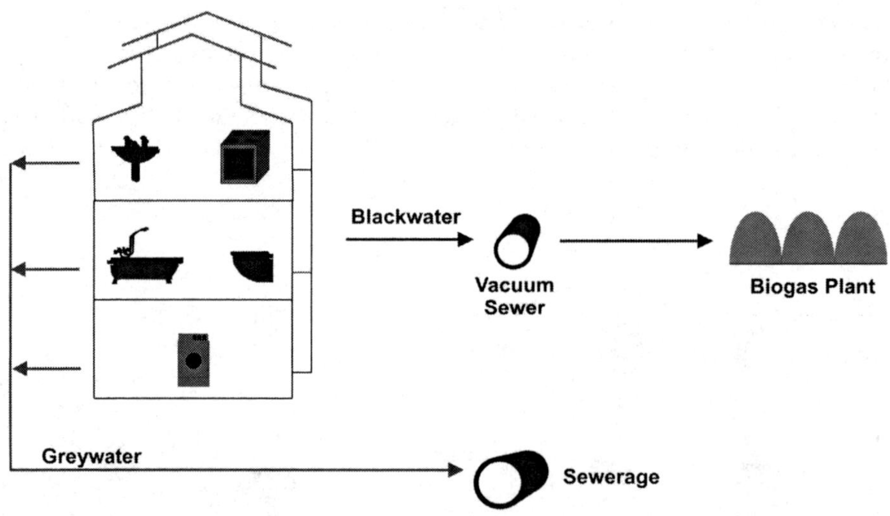

Fig. 8.3 System 3

for a sketch of the flows). In doing so, a wide scope of possible alternatives with varying degrees of connections to the existing waste water system is covered. Alternative systems, such as urine separation, aim at extracting valuable flows for further use, but can also reduce the pollution load in the centralized treatment plant, for example by reducing the incoming nitrogen load. Simulation runs have shown that the separation of for example urine significantly reduces the aeration requirements and has no limiting effect on nutrient availability for microorganism growth

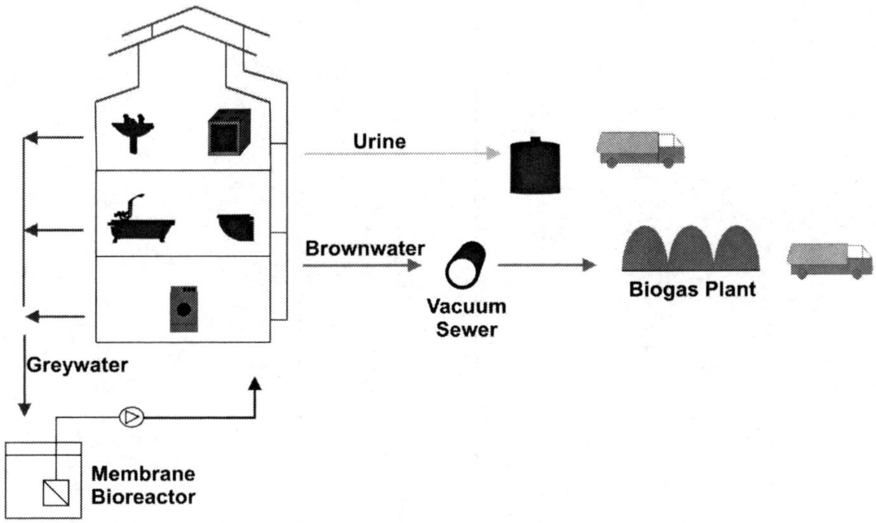

Fig. 8.4 System 4

in activated sludge reactors (Wilsenach and Van Loosdrecht 2003). One of the studied systems (system 4) even entirely eliminates the need for expensive centralized sewerage and treatment systems assuming that rainwater can also be used or discharged on-site. Of course, more variations and alternatives than the selected four systems are conceivable, but this was beyond the scope of this study.

Table 8.1 shows the four evaluated systems and highlights their respective treatment units as well as the varying degree of centralization. Systems 1 and 2 focus on the recovery of plant nutrients from urine as well as production of compost from faeces. Both systems reduce water consumption, since urine separating toilets require less water for flushing (depending on the toilet about 1–18 l liter per person and day) and composting toilets do not need any flushing water. Systems 3 and 4 introduce the aspect of energy recovery. Black water or brown water respectively are treated anaerobically, whereby methane and thus energy is produced. System 4 represents the most decentralized and self-sufficient system, including grey water recycling for the production of process and drinking water. At the same time, nutrients contained in urine are recovered through the use of urine separating toilets. Wherever conventional treatment is mentioned in table 8.1 it is assumed that the respective flows are treated in an already established manner. The interactions of conventional waste water treatment and urban form are not covered within the scope of this study.

8.3 Types of Urban Form

Since the aim of this study is to investigate the impact of certain characteristics of urban forms on the implementation of new sanitation concepts, a classification of existing residential areas according to their main characteristics is necessary.

Table 8.1 Treatment of source-separated waste water flows in considered systems

	Excreta/black water		
	Urine	Faeces	Greywater
System 1	Separation, nutrient recovery, use as fertilizer	Conventional treatment in central facilities	
System 2	Separation, nutrient recovery, use as fertilizer	Composting, use as soil conditioner	Conventional treatment (central)
System 3	Vacuum toilets, collection, anaerobic digestion in district facilities		Conventional treatment (central)
System 4	Separation, nutrient recovery, use as fertilizer	Vacuum toilets, collection, anaerobic digestion in district facilities	Decentralized treatment, use as process and drinking water

Different typologies are used in studies focusing on the interrelationship of urban form and infrastructure (e.g., Buchert et al. 2004; Pauleit 1998). For this study, a typology was developed based on other authors and own considerations with respect to the specific research question. The average characteristics of these nine types of urban form are shown in Table 8.2. Figures 8.5, 8.6, 8.7, 8.8, 8.9, 8.10, 8.11, 8.12, 8.13 sketch their spatial occurrences. Although not every possibility might be represented in this typology, it includes the major forms in the German context.

8.3.1 Selected Neighborhoods

To arrive at tangible results regarding the interaction of urban form and sanitation systems, existing neighbourhoods in Hamburg were identified as representative references for every type. These samples include unknown, archetypical neighbourhoods like the selected example for terraced houses in Hamburg-Langenbek, as well as neighborhoods, which are famous for their architectural qualities. These are for example the ensemble "Grindelhochhäuser" in Hamburg-Harvestehude, 12 slab blocks that have been built between 1949 and 1956. They are the first German housing project using steel frame construction techniques. Another example is the large scale development in Hamburg-Steilshoop, built in the 1970s and providing space for about 20,000 inhabitants. For more and detailed information regarding these neighbourhoods and their characteristics please refer to Ziedorn (2007).

Detailed data on the characteristics of the neighbourhoods is used to assess the general suitability of the four selected waste water systems. Furthermore, the treatment facilities can be roughly designed based on the available data in order to identify constraints and favourable conditions among the different housing types. For example, lengths of sewer pipes or collection rates for source-separated flows depend on physical characteristics such as housing or population density.

8.3.2 Socio-Economic Differences

In addition to the five physical characteristics mentioned above (see Table 8.2), also socio-economic data are available for the selected neighbourhoods. Factors such as age distribution, unemployment rate and size of households have an impact on waste water generation, thus, these factors should be considered when implementing new sanitation concepts into existing housing stock.

Unfortunately it is not possible to generalise these impacts. Certain correlations can be observed, but when it comes down to a specific neighbourhood it is necessary to evaluate the available data to specify the adjustments needed to realise a new sanitation project. The following two examples should give an idea of the possible impacts. Age distribution, unemployment rate, family size and related social features influence the timing, the volume and also the quality of the waste water production over

Table 8.2 Types of urban form and their average characteristics

Type	Floor-space index[a]	Number of storeys	Plot area in m^2	Buildings per ha.	Population density (inh./ha.)
1. Rural settlement	0.25	1.5–2	1,250	6–7	8–16
2. Detached and semi-detached houses	0.2–0.4	1.5–2	400	8–16	30–50
3. Urban mansions	0.6	2.5	750	8–12	40–60
4. Terraced houses	0.3–0.8	2	250	10–20	120–140
5. Perimeter block	1.2–3.5	3–4	350	8–16	260–320
6. Linear block	0.8–1.3	3–4	700	5–10	360–420
7. Slab block	1.6	10–15	11,500	~1	180–200
8. Large scale development	1.0–1.8	4–10	26,400	~0.5	120–150
9. Multistorey buildings	3.5–7	8–15	2,000	4–6	400–600

[a]Floor Space Index: ratio of the combined area of all storeys and the area of the plot

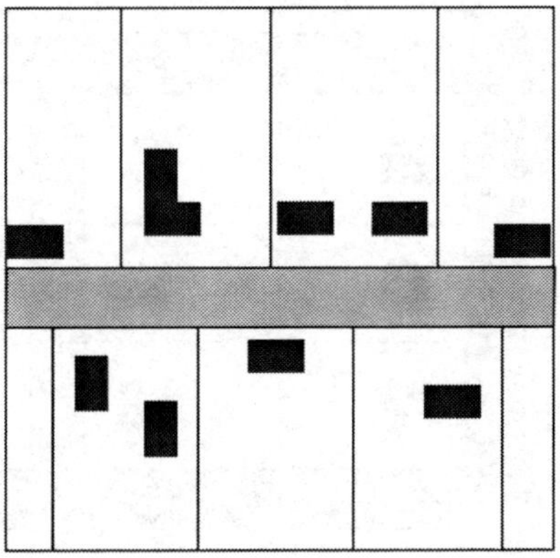

Fig. 8.5 Rural settlement

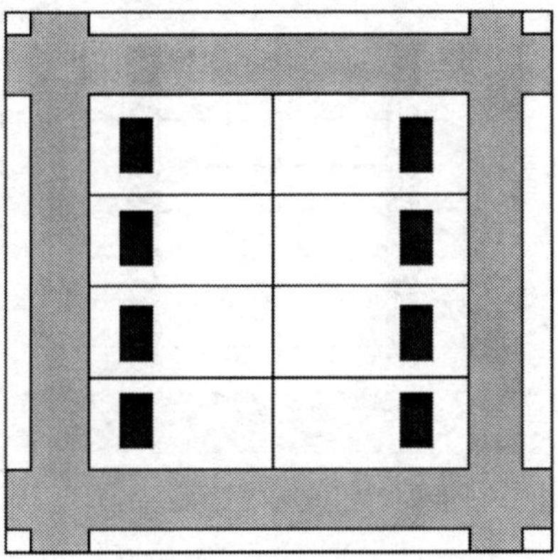

Fig. 8.6 Detached and semi-detached houses

the day. For example in a neighbourhood consisting of mostly younger, working singles the waste water generation generally has peaks in the morning and evening and is relatively low over the day. Contrary in neighbourhoods with a greater percentage of families and retired persons the generation is more equally distributed. This is due to the general presence of the inhabitants – children or older people tend to spend more time in their homes respectively their vicinity than younger persons,

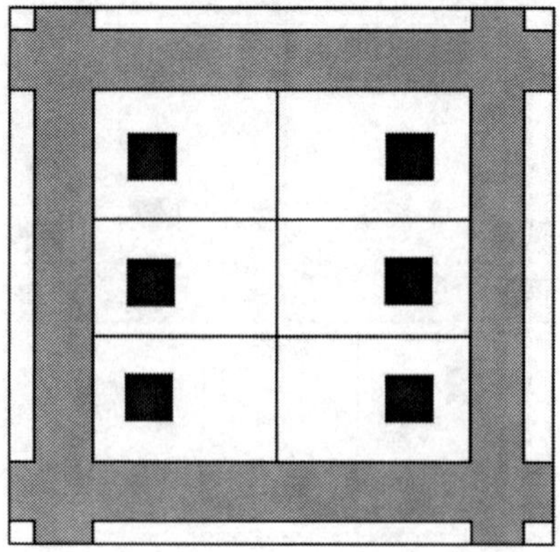

Fig. 8.7 Urban mansions

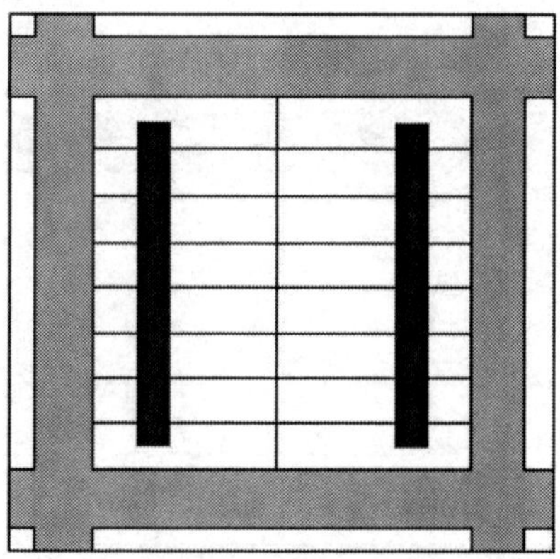

Fig. 8.8 Terraced houses

who work for example in other city districts and only stay home in the evenings or week-ends. A high percentage of elderly persons, who tend to consume more pharmaceuticals, can also affect the waste water characteristics (particularly of urine) due to the excretion of pharmaceutical residues.

Cultural and economic differences may be another aspect to be regarded. Water consumption, for example for dish washing or showering/bathing may differ depending on the cultural background or the ecological consciousness of the inhabitants. For example, Germans usually wash their dishes with relatively little water,

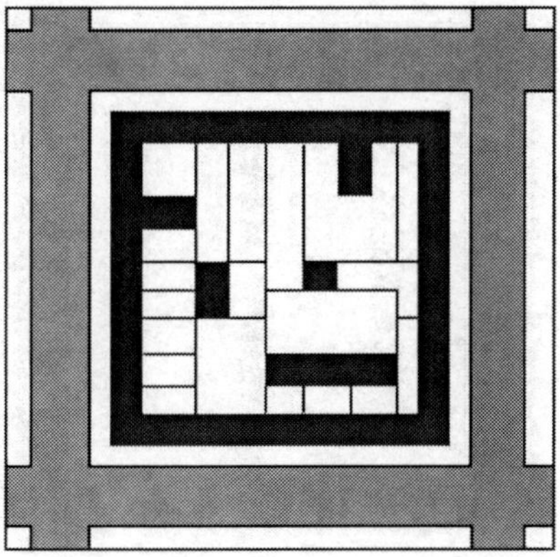

Fig. 8.9 Perimeter block

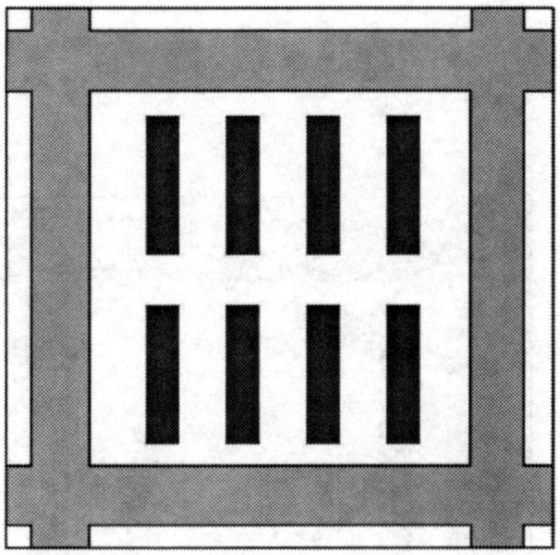

Fig. 8.10 Linear block

whereas other cultures prefer rinsing the dishes with running water. This affects the waste-water volume, but also the concentrations of pollutants. Economic constraints (e.g., high unemployment rates) resulting in the propensity to save can reduce household water consumption where the water and waste water fee is connected to water consumption. Refurbishments including the exchange of sanitary installations can have the same effect. This was observed in the East of Germany after the German reunification.

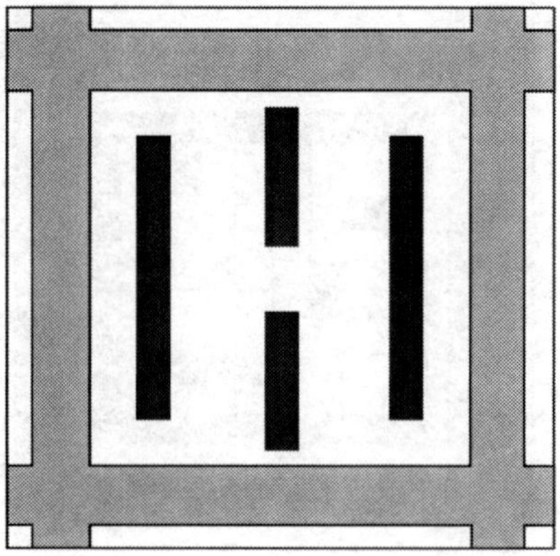

Fig. 8.11 Slab block

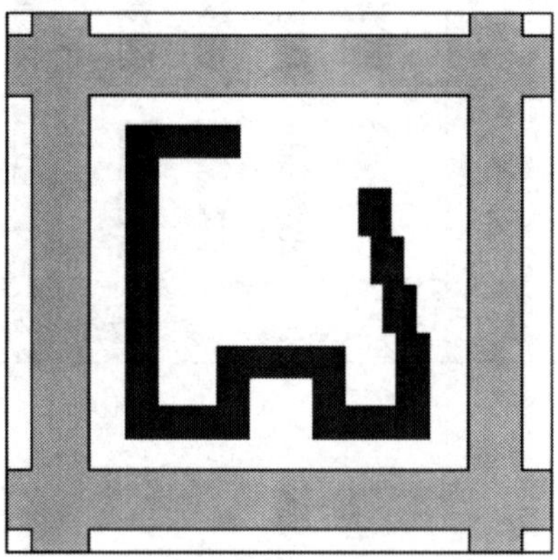

Fig. 8.12 Large scale development

Therefore, decentralized treatment of source-separated flows should be adapted to socio-economic factors prevalent in the neighbourhood under consideration. For example, enough storage capacity to buffer peak flows needs to be provided and treatment of the flows should be customized to the specific pollutant loads and concentrations.

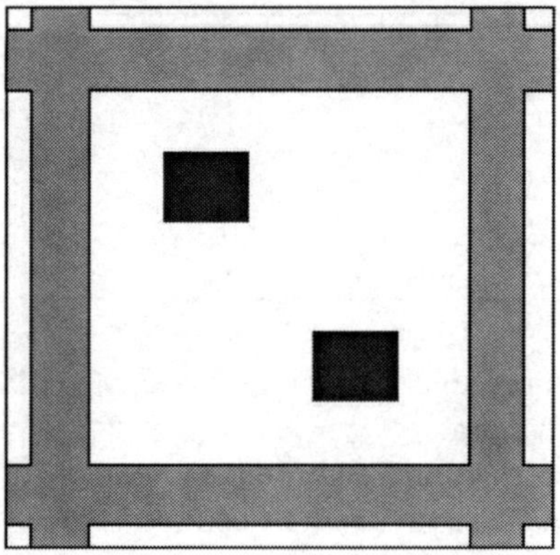

Fig. 8.13 Multi storey buildings

8.4 Basic Data and Assumptions on Waste Water Generation and Urban Form

In order to compare the different types of urban form, a detailed look is taken at the waste water generation and the resulting required treatment facilities and technical designs in the different selected neighbourhoods. Following general data on waste water generation rates are used (Table 8.3, based on Lange and Otterpohl 2000).

The urine is assumed to be collected in urine separation toilets and urinals (system 1, 2 and 4) and collected in containers with a volume of 1 m^3. The containers are assumed to be placed underground next to the street for emptying purposes. Vacuum trucks then can easily extract the urine by suction. For a detailed analysis two variants are assessed. Firstly, a fixed fortnightly collection cycle is assumed. This means that approximately 47 people can be connected to one urine collection tank. In buildings with more inhabitants, more containers or one with a volume larger than 1 m^3 needs to be provided. The urine from buildings with fewer inhabitants will be collected in a conjoint container. Collecting pipes will be required for this. The second variant is based on the assumption that every building is connected to (at least) one container with a volume of 1 m^3. The collection cycle is variable for every type of urban form and is based on the filling rate of the container, and thus the population density.

The collection of faeces without flushing water (system 2) is considered to occur within composters straight underneath the toilets. The installation of the composter requires additional space in the basement of the building. It is assumed that for five persons the required volume is 6.4 m^3 and required floor space is 3.1 m^2 (W. Berger 2007, Head of Berger Biotechnik). Up to now there is no experience with composting

Table 8.3 General data on waste water generation

Grey water	83	$l \cdot (cap \cdot day)^{-1}$
Urine	1.5	$l \cdot (cap \cdot day)^{-1}$
Faeces	0.15	$l \cdot (cap \cdot day)^{-1}$
Flushing water	3	l per flush (short)
	9	l per flush (full)
Flushing water (vacuum toilet)	1	l per flush

toilets in houses with more than four storeys due to ventilation requirements and risk of soiling of the downpipe.

For the vacuum sewers (system 3 and 4) a maximum of 1,500 inhabitants connected to one vacuum station is assumed and a maximum sewer line length of 2,000 m. In addition, the maximum connection rate is 0.2 persons per metre of vacuum sewer line. Concerning the anaerobic digester in system 4, the required floor space is assumed to be 100 m^2 including distance space to neighbouring buildings. The digesters are assumed to serve about 350 inhabitants.

8.5 General Suitability Check of the Systems

A first qualitative assessment of the suitability of the systems for the different types of urban form results in the exclusion of some of the systems for particular types (see Table 8.4). This is due to either the lack of space for installations of decentralized treatment plants (in particular biogas plants) or to the height of the buildings making them unsuitable for the installation of dry faeces collection in composters.

8.5.1 System 1

Generally, there are no particular constraints from a spatial point of view for the implementation of urine separation toilets and urine collection tanks. Open space in yards or basements is required for the installation of the urine collection tanks. Housing layout and population density affects the number of tanks and collection cycles. With increasing population density the number of required tanks increases, while the length of required sewer line decreases. Retrofit-ting buildings with urine separation requires the installation of new toilets as well as separate urine pipes.

8.5.2 System 2

For urine separation the same as above (system 1) applies. Regarding the collection and treatment of faeces in composters, the main constraint is the number of storeys. Due to that, only housing types 1–4 seem suitable, since in those houses the number

Table 8.4 General suitability of the sanitation systems

Type	System 1	System 2	System 3	System 4
1	+	+	+	+
2	+	+	+	0 (S)
3	+	+	+	0 (S)
4	+	+	+	–
5	+	0 (H)	+	0 (S)
6	+	0 (H)	+	0 (S)
7	+	–	+	0 (S)
8	+	–	+	0 (S)
9	+	–	+	+

+ = suitable
– = not suitable
0 = limited suitability depending on space (S) or height of buildings (H)

of storeys is usually less than four. However, the composting system could be modified to a dehydration system based on vaults with regular collection (e.g., every 6 months). Such dehydration vaults could be installed in every storey superseding the long and problematic shafts. All in all, the replacement of flush toilets with a dry system requires space in the buildings, which might not be available. In addition, the retrofitting would need structural modifications that are difficult to implement on a large scale. Therefore, dry collection of faeces seems feasible rather in new development areas than in existing housing stock.

8.5.3 System 3

System 3 requires the retrofitting of vacuum toilets and vacuum pipes in the buildings. This should be done preferably during general refurbishment works. An example of such a retrofitting with vacuum toilets is the equipment of a building with 32 flats constructed 35 years ago in Hanover, Germany (Herrmann 2003). Vacuum stations and anaerobic digesters need to be accommodated within the neighbourhood. The availability of communal areas can play a decisive role for the selection of possible locations. The housing layout and the population density affect the vacuum pipe network.

8.5.4 System 4

Similar aspects of systems 1 and 3 apply to system 4, which combines vacuum transport and anaerobic digestion of brownwater with urine separation. The more decentralized treatment of brownwater (as described above) requires availability of space for the implementation of biogas plants and, for example, combined heat and power plants within the neighbourhoods. This space is generally available in the studied neighbourhoods, but only on private plots. If only municipal plots are to be used, restrictions concerning several types of urban form occur (as shown in Table 3). In addition, greywater treatment in, for example, bio-membrane reactors with subsequent disinfection takes up space within the buildings.

8.6 Cost Assessment for Urine Separation

Urban form generally has an effect on the costs of water-related infrastructure. For example, Siedentop et al. (2006) showed that decreasing housing density results in increased pipe requirements and, thus, increased costs in relation to floor-space.

To exemplify the impact of different housing types on implementation costs, an assessment of the installation of urine separation in the nine selected neighbourhoods is carried out. As described above two variants are evaluated. The first comprises a fixed collection cycle of 14 days and joint collection tanks wherever necessary. The second refers to a variable collection cycle and 1 m^3 containers for every individual building. For the second variant the cost of collection pipes decreases, but the investment costs for urine collection tanks are higher. In addition, as a result of the longer collection cycle, costs for collection by vacuum trucks will occur more rarely. However, operational costs are not included in this analysis due to the scope of this study and therefore logistical requirements would need to be more thoroughly examined. In general, specific collection costs will decrease in more densely populated areas, since transport distances decrease.

The pipe requirements are calculated based on following assumptions: The number of inhabitants per building can be used to check if and how many buildings need to be connected to a joint tank in variant 1. Wherever this is the case, the length of the connecting pipe is calculated from the size of the respective plots. The required length of the urine collection pipe to the tank on the plot (private house connection) is based on a relationship shown by Thoma and Goetz (2008) developed for a project on house connection pipes on private premises in Germany. The length of pipes inside the buildings is based on the assumption that about 5.8 m are required per dwelling (Oldenburg and Dlabacs 2007). For an overview of the installation requirements please refer to Table 8.5, which is based on data from the selected reference neighborhoods.

It is assumed that in every house (i.e. housing types 1–4) two toilets are installed per dwelling, respectively in multi-family houses one toilet per apartment. The costs of toilets and urine pipes including installation are based on available cost figures for retrofitting a building with urine separation (Oldenburg and Dlabacs 2007). Particularly when retrofitting of additional pipes is done within the scope of general sanitary refurbishments, costs differ only slightly from new installations (Kaiser 2008). The following cost estimates are used for the assessment (Table 8.6).

Figure 8.14 illustrates the resulting specific costs for the nine types of urban form. Please note that the investment costs shown here do not reflect absolute values, but only the differences between the housing types. Those types with lower densities (particularly types 1–3) have considerably higher specific installation costs of urine separation. This is a result of higher house connection costs due to

Table 8.5 Installation requirements

Type of Urban Form	1	2	3	4	5	6	7	8	9
Ratio of built area to site area	0.16	0.32	0.22	0.4	0.65	0.43	0.09	0.5	0.21
Dwellers per building	4	2	6	12	14	60	250	864	105
House connection per person (m.)	19	18	9	1	6	5	2	4	3
Pipes inside the building per person (m.)	2.5	1.5	1.4	1.1	2.3	1.5	2.6	1.2	1.9
Variant 1									
No. of containers per building	NA	NA	NA	NA	NA	2	6	19	3
No. of buildings per container	13	24	8	5	4	NA	NA	NA	NA
Length of connection pipes per person (m.)	10	10	5	1	1	NA	NA	NA	NA
Variant 2									
No. of containers per building	1	1	1	1	1	2	6	19	3

Table 8.6 Costs for urine separation systems

Retrofitting items	Costs
Urine pipe (house connection on premises) DN 150	45 €m^{-1}
Urine pipe (inside the building) DN 70	15 €m^{-1}
Urine pipe (connection pipe) DN 150	104 €m^{-1}
Urine separation toilet	450 €
Urine storage (1 m^3)	1,500 €

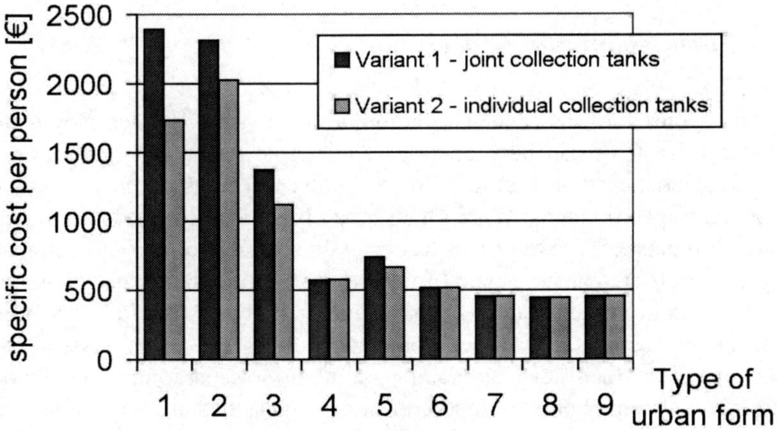

Fig. 8.14 Specific costs for the nine types of urban form

larger specific plot sizes. Another reason is the assumed equipment with two urine-diversion toilets, which are currently rather costly.

8.7 Factors Affecting the Ease of Implementation

The development of the typology of urban form and subsequent assessment of the different types shows that a variety of factors affect the planning and implementation of new sanitation concepts.

8.7.1 Physical Aspects

The following physical or spatial factors are identified as having the most significant impact on the retrofitting of source-separating sanitation concepts:

- Population density: As shown in the cost assessment of a urine separation system, the population density affects hardware requirements. A higher density can result in lower specific costs per person.
- Housing layout: The layout influences the options for joint systems and pipe requirements. For example, a more compact housing layout requires fewer pipes than a more elongated one. In addition, the feasibility of combined treatment facilities for neighboring houses depends on the general housing layout.
- Ratio of built-area and open space: Source-separating systems relying on decentralized collection and treatment installations require available space. Particularly in dense areas there is a high demand and competition for the use of open space.
- Number of storeys: The number of storeys generally affects the population density. In addition, little experience is available on the implementation of some source-separating technologies (particularly dry toilets) in multi-storey buildings.

8.7.2 Socio-economic Aspects

Factors like household size, age distribution of the inhabitants, unemployment rates and cultural practices can have an impact on waste water generation and waste water characteristics. For example, these factors affect time spent at home and therefore also specific loads. Within the scope of this study, these aspects were not considered in detail, since the focus has been on spatial and physical parameters.

Social aspects are hard to generalise and more specific to the special conditions. When it comes to implementation, planners and engineers should be aware of the importance of such factors. Since source-separating sanitation systems require 'proper' use of the facilities (for example, some urine separating toilets work only if users sit), awareness creation and educational measures are crucial for the successful implementation of new sanitation concepts in existing residential areas. In new developments there is often a greater awareness and willingness to go beyond conventional solutions. In existing housing areas, however, long established habits can prove it to be more difficult in creating acceptance of new technologies.

8.8 Conclusion

Retrofitting of source-separating technologies into existing housing stock is possible, resulting in alternative sanitation systems aiming at higher resource efficiency. The type of urban form has an impact on planning and implementing source-separating sanitation systems. Therefore, a good coordination and collaboration of urban planners and waste water professionals is the basis for a successful transition of the urban sanitation system. The analysis of urban form and existing housing stocks can be useful for the identification of areas where a greater benefit-cost-ratio can be anticipated or where potential hindrances for implementation exist. However, for the selection of areas where transition is supposed to take place, also other factors like refurbishment requirements or current urban (re)development projects will be crucial. In addition, more research is needed in the study of social, cultural and logistical aspects and their dependency on types of urban form. This will have a direct impact on a cost comparison including operational costs.

References

Buchert, L., Deilmann, M., Fritsche, U., Jenseit, W., Lipkow, A., Rausch, C., et al. (2004). Nachhaltiges Bauen und Wohnen in Deutschland (Sustainable Con-struction and Housing in Germany. In German). Schriftenreihe Texte des Umweltbundesamtes 01/04. UBA, Berlin, Germany.

Herrmann, T. (2003). Blackwater separation in a residential house by vacuum toilets. In C. Maksimovic, D. Butler, & F. A. Memon (Eds.), *Advances in water supply management, proceedings of the international conference on computing and control for the water industry* (pp. 507–516). London: Swets & Zeitlinger, September 15–17, 2003.

Herzer, P. (2004). *Einflüsse einer naturnahen Regenwasserbewirtschaftung auf den Städtebau – Räumliche, ökonomische und ökologische Aspekte* (Impacts of a nature-orientated Rainwater Management on the Urban Development – Spatial, Economical and Ecological Aspects. In German). Stuttgart, Germany: IRB Buch.

Kaiser, M. (2008). Regenwasserbewirtschaftung in Kombination mit Regenwassernutzung im Bestand (Rainwater Management in Combination with Rainwater Use in Existing Housing Stocks. In German). In: fbr – Wasserspiegel. Fachvereinigung Betriebs- und Regenwassernutzung e.V. (fbr). (Vol. 1(8), pp. 3–5).

Lange, J., & Otterpohl, R. (2000). Abwasser – Handbuch zu einer zukunftsfähigen Wasserwirtschaft (Wastewater – Handbook for Sustainable Water Management. In German), MALL-BETON GmbH, Donaueschingen-Pfohren, Germany.

Löber, T. (2001). Beitrag zu einer städtebaulich neuorientierten Regenwasserbehandlung in Wohnsiedlungen (New Approaches to Rainwater Management in Built-Up Areas. In German). Ph.D. thesis, Hochschule der Künste, Berlin, Germany.

Oldenburg, M., & Dlabacs, C. (2007). Final cost calculation report for the demonstration project "Sanitation Concepts for Separate Treatment of urine, faeces and greywater" (SCST). Kompetenzzentrum Wasser Berlin gGmbH.

Otterpohl, R. (2001). Design of highly efficient source control sanitation and practical experiences. In P. Lens, G. Zeeman & G. Lettinga (Eds.), *Decentralised sanitation and reuse: Concepts, systems and implementation*. London: IWA Publishing.

Pauleit, S. (1998). Das Umweltgefüge städtischer Siedlungsstrukturen (Environmental Impact of Urban Form. In German). Ph.D. thesis, Technische Universität München, Germany.

Schäfer, D., & Rudolph, K.-U. (2001). International Survey on Alternative Water Systems. BMBF-Research Project No. 02 WA 0074. Federal Ministry of Education and Research, Bonn – Karlsruhe – Witten, Germany. Retrieved February 11, 2008, from http://uni-wh-utm.de/download/aws-engl.pdf

Siedentop, S., Schiller, G., Koziol, M., Walther, J., & Gutsche, J.-M. (2006). Siedlungsentwicklung und Infrastrukturfolgekosten – Bilanzierung und Strategieentwicklung (Urban Development and Infrastructure Costs – Balancing of Accounts and Development of Strategies. In German). BBR-Online-Publikation Nr. 3/2006. Retrieved: August 28, 2006, from http://www.bbr.bund.de/cln_005/nn_21272/DE/Veroeffentlichungen/BBR-Online/BBR-Online.html

Thoma, R., & Goetz, D. (2008). Zustand von Grundstücksentwässerungsanlagen (Status of Private Sewerage Systems. In German). In: *KA – Abwasser, Abfall* (Vol. 55(2), pp. 116–130).

Wilsenach, J. A., & van Loosdrecht, M. C. M. (2003). Impact of separate urine collection on wastewater treatment systems. In: *Water science & technology* (Vol. 48(1), pp. 103–110). UK: IWA Publishing.

Ziedorn, V. (2007). Städtische Siedlungsstrukturen und dezentrale Abwassersysteme (Urban Form and Decentralised Wastewater Systems. In German), HafenCity Universität, Hamburg, Germany.

Zink, U. (2007). Sanierungsbedarf im Gebäudebestand (Refurbishment Demand in Existing Housing Stock. In German). In: Nachrüstung von innovativen Wasserkonzepten in Gebäuden. 27.11.2007. fbr (ed.).

Chapter 9
Reconsidering Urban Sewer and Treatment Facilities in East Africa as Interplay of Flows, Networks and Spaces

Sammy Letema, Bas van Vliet and Jules B. van Lier

Abstract Urbanization has brought about concentrations of people in densely populated settlements, resulting in the generation of waste water that needs to be disposed off in a hygienic way to avoid the outbreak of diseases. Decisions on what area to sewer, the nature of sewer schemes and treatment works to be used, and the kind of collection and transport system to adopt is often complex and difficult to make. This chapter (re)considers urban sewers and treatment works as the interplay of flows, networks and spaces, and puts forward a conceptual framework for decision-making. It examines current and future sanitation structures in Kampala and Kisumu in terms of sanitation flows, sanitary networks and demands for space. Knowledge of such sanitation structures serves as an input to the assessment of opportunities for so-called Modernized Mixtures of sanitation systems in cities around Lake Victoria.

9.1 Introduction

Traditionally in East African cities, the icon of sanitation modernization was indisputably the construction of a large centralized sewerage network complemented with 'Western' conventional treatment works consisting of mechanized aerobic treatment

S. Letema (✉)
Department of Environmental Planning and Management, Kenyatta University, Nairobi, Kenya
e-mail: sletema@yahoo.com

B. van Vliet
Environmental Policy Group, Wageningen University, Hollandseweg 1, 6706 KN Wageningen, The Netherlands
e-mail: bas.vanvliet@wur.nl

J.B. van Lier
Department of Water Management, Section Sanitary Engineering, University of Technology, Stevinweg 1, Delft, The Netherlands
e-mail: j.b.vanlier@tudelft.nl

B. van Vliet et al. (eds.), *Social Perspectives on the Sanitation Challenge*,
DOI 10.1007/978-90-481-3721-3_9, © Springer Science+Business Media B.V. 2010

systems in an effort to modernize the city and protect public health (Nilsson 2006). However, over the years, sanitary modernization has generally resulted in three configurations: centralized systems in the planned urban centres; decentralized catchment systems in incremental and ad hoc planned urban areas, satellite systems in suburban planned settlements, and on-site technologies in peri-urban areas. About 70% of East African city population lives in informal and slum settlements beyond the reach of modern sewer networks. Such settlements are considered traditional within the discourse of urban planning and modernization, and as such, often ignored, labeled illegal or torn down in the name of modernization. Implementation of centralized systems and on-site technologies in East African cities has been viewed as unsustainable in the long run. For instance, Spaargaren et al. (2005) note that implementation of centralized systems has been less successful, has reinforced inequality, bleeds money out of the social system, and runs counter to local sustainability, whereas alternative household on-site technologies based on percolation properties are challenged by rapid urbanization and high demand for safety, accessibility and eco-efficiency. Therefore, improved on-site sanitation, although mostly applied, is not considered as integrated solutions in cities, since it is limited to only mitigating the environmental and health nuisance. There is a shift towards the implementation of catchment and satellite based sewerage and the upgrading of existing on-site technologies into sewers in order to avoid polluting the environment and to ensure public health. However, criteria on what area to sewer, on the nature of sewer schemes to be used, and on the kind of treatment works to adopt need to be applied in the context of a decision-making framework which fits into the local conditions of existing East African cities. Therefore, we propose to base the critical threshold for the adoption of particular sanitation systems and technologies in a particular area and at a particular scale (i.e. city, catchment area, neighborhood or estate, cluster, blocks, and household) on a set of key criteria referring to the sanitary flows, their networks and their demands for space primarily.

This chapter explores the interplay of sanitary flows, networks and space demands as criteria for determining the boundary conditions and decision-making at strategic city level for the development of diverse patterns of sanitation systems in East African cities taking Kampala (Uganda) and Kisumu (Kenya) as its main cases. The findings are based on methods and concepts presented in the following Section 9.2. In the three sections to follow, respectively the sanitary flows, the networks and the spaces in Kampala and Kisumu are being discussed, leading to an analysis and discussion on sanitation reconfiguration criteria for selecting sewerage areas and systems (Section 9.6). We conclude in Section 9.7 with a discussion on the use of the Modernized Mixture Approach for analyzing and planning sanitation in East African urban centres.

9.2 Conceptual Framework and Methodology

The conceptual framework within which sanitary systems in Kampala and Kisumu are analyzed is shown in Fig. 9.1. Three typical sanitary configurations can be analyzed by the flows they convey, the networks they encompass and the spaces they occupy.

In this chapter the waste water flows generated per day per hectare (base flow density – m^3/day/ha); waste water discharge quality; extent and nature of network

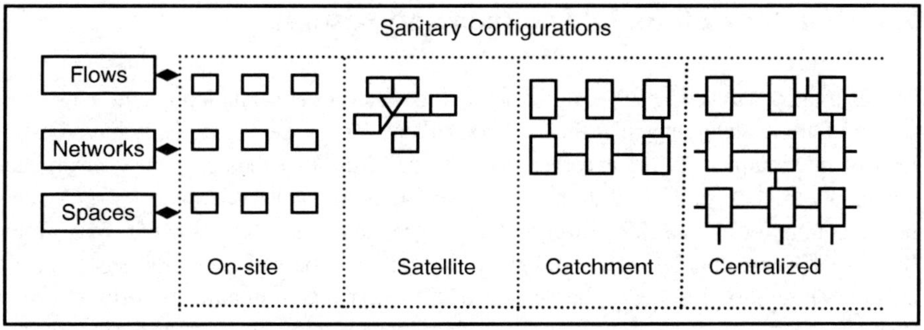

Fig. 9.1 Sanitary configurations in East African cities

connections and mechanization; sanitary space characteristics and population equivalent per hectare (p.e./ha) are used to establish the feasibility for different sewage systems. Besides it incorporates a socio-technical orientation on flows, networks and spaces, as it considers demographic, institutional and social aspects in discussing the current and planned sewage systems in Kampala and Kisumu.

Generally, domestic sewage flows can be categorized into various flowstreams such as grey, yellow, brown and black water, but in East African cities in the short and medium-term, waste streams will mostly be combined and called domestic sewage, except for on-site sanitation systems. However, domestic sewage and storm water shall be separated as is the current design practice.

Waste water flows estimates for sewage are based on water consumption levels with the assumption that 80% of the consumed water is discharged as waste water derived from the technical reports. Base flow density is begged on attaining sufficient concentration of medium and high income water users generating sufficient sewage flows for sewers to function and run the treatment works sustainably. In Kampala city, the waste water flow density for sewage is 10 m^3/day/ha, equivalent to 50% of households in an area with a population density of 200 p.e./ha having an in-house water connection (NWSC 2004a, b). In Kisumu the critical density for sewerage is 120 p.e./ha based on population estimates (MWI 2008). However, for uniformity, base flow density is computed for Kisumu to be in tandem with Kampala.

Our findings are based on studies of historical operation and maintenance records, technical reports,[1] field surveys and expert interviews.[2] The research was conducted between October 2007 and December 2008. Sanitation regulations and standards were evaluated in order to establish the discharge standards and thus determining the appropriate treatment technology process.

[1] Kampala Sanitation Strategy and Master Plan (Volume 1 & 2, and Appendices); Kampala Sanitation Program (Feasibility Study); Kisumu Water Supply and Sanitation Project (Sewerage Design Report and Feasibility Report); Practice Manual for Sewerage and Sanitation Services in Kenya; and KIWASCO Sanitation Sector Investment Plan, 2008.

[2] City planners, Kisumu Water and Sewerage Company (KIWASCO) and Kampala Water Partnership network managers; sewerage analyst; superintendents of works and zonal managers. Moreover, a Lake Victoria South Water Services Board (LVSWSB) technical services manager; asset development manager and water quality analyst were interviewed.

9.3 Sanitary Flows in Kampala and Kisumu

The estimated waste water flows in Kampala to waterborne sanitation (septic tanks and sewers) for all catchments in 2008 is 54,000 m^3/day. The sewerage capacity has a design load of maximum daily dry weather flow of 15,000 m^3/day thus can only convey and potentially treat about 23% of the flows. The population connected to waterborne sanitation accounts for about 6.5% sewer and 17.5% septic tanks. The estimated waste water flows in Kisumu in 2007 was 34,000 m^3/day whereas on-site sanitation systems discharged about 3,600 m^3/day. The design load of the existing treatment works can only treat 17,800 m^3/day of the flows. However, the flows reaching public treatment works are 11,000 m^3/day (Table 9.1) out of which only 9,800 m^3/day can be treated, since the Kisat STW design load has already been surpassed by 32% while Nyalenda ponds receives about 30% of design load. Kisumu Molasses treatment works is the only functional satellite system in Kisumu city, treating industrial waste water. The other industrial pre-treatment plants (Kisumu Cotton Mills and Kenya Breweries) are no longer in operation.

The main treatment works in Kampala (Bugolobi) and Kisumu (Kisat) are already overloaded with flows. In Kisumu, flows may be redirected by aid of pumping to Nyalenda waste stabilization ponds to utilize existing capacity. In Kampala, pumping the sewage flows to Ntinda (WSP), Naalya and UNISE satellite ponds is not feasible, because firstly, they are in different drainage catchments and thus very expensive to maintain. Secondly, they are developed by different private and semi-public institutions for specific urban clientele. Thirdly, it has different management and organizational structures that are not commercially oriented unlike the National Water and Sewerage Corporation (NWSC) systems.

Kampala and Kisumu city use waste water treatment by conventional attached growth STW and Waste Stabilization Ponds (WSP). However, the effluents are not

Table 9.1 Flows load and capacity in Kampala and Kisumu (NWSC 2004b; LVSWSB 2008)

Sewage treatment works	Existing load (m^3/day)	Design load (m^3/day)	Available (m^3/day)	Sewer type
Kampala City				
Bugolobi Conventional STW	8,907	8,907	0	Centralized
Bugolobi East WSP	365	187	0	Satellite
Kyambogo WSP	230	230	0	Satellite
UNISE WSP	53	360	247	Satellite
Luzira Prison WSP	270	–	–	Satellite
Naguru WSP	–	–	–	Disused
Naalya WSP	460	1,902	1,442	Satellite
Namboole Stadium WSP	155	155	0	Satellite
Ntinda WSP	206	439	233	Satellite
Kisumu City				
Kisat STW	9,000	6,800	0	Centralized
Nyalenda WSP	3,000*	11,000	8,000*	Catchment
Molasses WSP	800	–	–	Satellite

*Data obtained from field survey by Author in 2008

Table 9.2 Treatment mix process options based on discharge standards

	Discharge parameters removal				
Technology process	Carbon (C)	Faecal coliform (FC)	Nutrients (N)	C + FC	C + FC + N
PST-TF	++	–	–	+	+
PST-TF-M	++	++	+	++	+
PST-TF-M-W	++	++	++	++	++
UASB-F	++	–	–	+	+
UASB-F-M	++	++	+	++	+
UASB-F-M-W	++	++	++	++	++
A-F-M	++	++	+	++	+
F-M	++	++	++	++	+
A-F-W	++	+	++	+	+
F-W	++	+	++	+	+

Symbols: Relative removal levels are denoted by – (minimal), + (average), ++ (high); A (Anaerobic), F (Facultative), M (Maturation), PST (Primary Sedimentation Tanks), TF (Trickling Filter), UASB (Up-flow Anaerobic Sludge Bed), W (Wetland [existing 16.9 ha])

complying with National Environment Management Authority (NEMA) standards. Previous studies, for instance, showed that performance of treatment systems are not in tandem with environmental discharge standards for BOD, COD, TSS, P and N (NWSC 2004b). Apparently, the adopted treatment technology, conventional and waste stabilization ponds, are inappropriate technologies for the investigated conditions. To meet discharge standards at least operation and maintenance costs, a shift from the traditional approach to flexible mixtures to treatment works is imperative (Table 9.2). Combination of UASB-TF that is not evaluated above is an improvement of PST-TF. Therefore, treatment technology combinations if designed, managed and operated well, will work also in East Africa. The prevailing conditions are apparently limiting its application due to slow technological adoption and personnel capacity. The incorporation of P and N in discharge standards, moreover, makes use of conventional techniques in developing countries much more expensive.

The flows density in a particular area can be used to assess technical feasibility for sewerage extensions, establishment of new small and medium-large scale sewerage schemes. Using base flow and population density, lowest administrative tiers, that is parish in Kampala and sub-locations in Kisumu, are used to demonstrate areas where threshold values have been attained for sewerage developments.

9.3.1 Configurations in the Making: Base Flow and Population Density as Determinants of Sewerage Development

Sewerage coverage in Kampala and Kisumu cities are 6.5% and 28.9% of the population, respectively. The organic and informal nature of development makes it difficult to distinguish areas that have attained threshold levels for sewerage, based on land use classification and planning standards. Consequently, the use of population and base flow density provides an alternative approach for determining such areas. Parishes and sub-locations outside the existing sewerage and city centre are used for Kampala and Kisumu respectively, to depict how population and base flow densities can guide sewerage development in phases.

9.3.1.1 Kampala City

Kampala's public-driven sewerage developments can be categorized into catchments, that is Nakivubo catchment extensions and proposed sewerage catchments (Table 9.2). The envisaged development of Nakivubo treatment works will necessitate extension of sewerage networks into surrounding parishes within Nakivubo catchment. The criteria for extension are based on base flow and population density. The parishes that have relatively high waste water flows and population density and thus meet the base flow density for sewerage extensions by 2013 are Bukesa, Namirembe, Kibuli, Kabagala, and Mengo (Table 9.3).

The proposed 2023 sewerage extensions will cover Kiswa, Katwe II, Wabigalo, Kisugo and part of Bukasa parishes. Kiswa extension is an exception since much of the flow is expected to arise from institutional developments. Although Kiswa shows low flow levels at the moment. The base flow is expected to rise since developments are often spatially concentrated, with large parts of land left for recreational and circulation land uses. The sewerage extension 1, which will serve parts of Naguro 1 and Nakawa parishes, will attain the base flow density for sewerage well beyond 2033. The flows and population densities will still be low enough for septic tanks to operate effectively until 2023. However, the area is predicted to be characterized by institutions and planned medium and high income residential areas, which usually show high demand for sewerage services. Hence, the construction of the extension may occur by 2023 (NWSC 2004b).

None of the new sewerage catchments – Lubigi, Nalukolongo and Kinawataka meets the criteria for provision of sewerage services until 2023 and in the case of Nalukolongo and Lubigi sewerage would not be required until 2033. However, parishes like Wandegeya in Lubigi catchment, and Kibuye II, Makindiye II, and Najjanankumbi in Nalukolongo catchments are best candidates for small-scale satellite systems development that can be integrated into catchment sewerage systems when it is developed later on. The development of catchment based sewerage schemes can be phased within and between the catchments. For instance, Kinawataka catchment may be scheduled for 2023 while Nalukolongo and Lubigi catchments are scheduled for 2033 and beyond. The base flow density will allow construction of the Kinawataka extension in phases. The Trunk sewer, Mbuya II (south), Mutungo (north), Mutungo (south) and the STW can be scheduled for 2023. Sewerage of the Ntinda (east), Naguro II, Ntinda (west), Mbuya II (north) sewer lines may follow at a later stage depending on urban and demographic developments.

Table 9.3 Population and base flow density in Kampala city (NWSC 2004b)

	Population 2033		Base flow density (m^3/day/ha)				
Parish	Density (p.e./ha)	Size(ha)	2013	2023	2033	Area (ha)	Flow 2033 (m^3/d)
Nakivubo sewerage catchment extensions							
Naguro 1	26.3	2,141	1.77	1.82	1.96	81	242
Nakawa	107.5	6,731	4.71	5.12	6.31	63	612
Bukesa	149.2	9,336	8.99	***9.64***[a]	**11.23**[b]	63	1,092
Nakasero 3	32.4	693	6.38	6.52	7.27	21	241
Kiswa	93.4	2,995	3.10	3.64	4.93	59	242
Namirembe	*201.4*[c]	5,992	***9.79***	**10.47**	**12.21**	30	562
Kibuli	241.3	8,306	***9.74***	**9.92**	**10.93**	34	582
Kabalagala	192.5	5,426	6.08	7.61	9.94	28	434
Mengo	*201.4*	4,059	***9.79***	**10.47**	**12.21**	20	381
Katwe II	*263.2*	8.312	7.04	7.76	**10.44**	32	510
Wabigalo	*225.5*	5,495	8.28	9.15	**10.73**	24	314
Kisugu	*257.2*	30,206	8.69	**10.90**	**14.25**	117	2,008
Bukasa	119.4	3,482	4.49	5.52	6.69	29	234
Lubigi sewerage catchment							
Bukoto I	187.3	32,774	5.33	7.80	**10.88**	175	2,285
Bwaise I	*286.9*	37,854	6.85	8.92	**11.11**	132	1,759
Kyebando	*257.6*	72,283	5.40	8.94	**14.02**	281	4,721
Makerere I	129.3	17,012	4.26	4.73	5.30	132	837
Makerere University	69.2	4,238	4.76	5.76	6.39	61	539
Wandegeya	183.9	5,171	8.03	**10.89**	13.89	28	469
Nalukolongo sewerage catchment							
Kabowa	*250.5*	34,379	6.82	***9.20***	**12.10**	137	1,993
Kibuye II	*203.7*	4,720	8.81	***9.87***	**10.23**	23	284
Makindye II	190.9	7,688	6.00	8.62	**11.28**	40	545
Najjanankumbi I	*213.5*	12,206	6.04	8.62	**11.66**	57	800
Najjanankumbi II	134.1	8,525	4.77	6.33	8.02	64	612
Ndeeba	101.2	7,988	2.47	3.49	4.86	79	460
Kinawataka new sewerage catchment							
Mbuya I	*257.6*	12,030	7.87	**11.44**	**15.37**	47	861
Mbuya II	99.5	15,068	3.45	4.43	5.37	151	976
Mutungo	*284.0*	54,170	8.88	**11.05**	**14.83**	191	3,394
Ntinda I	54.7	18,482	7.97	7.73	8.22	338	3,333
Naguro I	–	–	1.77	1.82	1.96	–	–
Naguro II	–	–	8.40	**11.84**	**18.66**	–	–

[a] The numbers in ***bold and italic*** indicate the areas that are at verge of base flow threshold level
[b] The numbers in **bold**[3] indicate the areas that have attained/surpassed sewerage base flow density threshold
[c] The numbers *italic*[4] indicate the areas that have attained requisite population density (p.e./ha) for sewerage

[3] The base flow density threshold for sewerage is 10 m^3/day/ha as adopted in the Kampala Sanitation Strategy and Master Plan 2004.

[4] The population density threshold for sewerage is 200 p.e./ha as adopted in the Kampala Sanitation Strategy and Master Plan 2004.

9.3.1.2 Kisumu City

Sewerage network in Kisumu covers about 10% of the land area and 28.9% of the population. Further sewerage development not only needs to be cost-effective but also operated and managed sustainably, which can be determined by the population and base flow densities across each sub-location of Kisumu city (Table 9.4). The table shows that already Kibuye, Nyalenda and Manyatta sub-locations have attained the required base flow and population density threshold for sewerage by 2007. Kibuye sub-location is dismally connected with Migosi having only 30 active sewer connections out of 1,371 possible connections. Despite Nyalenda and Manyatta surpassing the threshold significantly, in the former there are no sewer connections whereas in the latter only 94 connections are active despite the eastern trunk sewer passing through the zone. Some sub-locations that have very low population and base flow densities are sewered such as Milimani with 44.8 p.e/ha and 1,270 sewer connections and Wathorego with 16.2 p.e./ha and 70 sewer connections.[5]

Table 9.4 Population and base flow density in Kisumu city[6]

	Population 2030		Base flow density (m³/day/ha)				
Sub-location	Density (p.e./ha)	Size area (ha)	2007	2010	2020	2030	Flow 2030 m³/day
Kibuye	*155.8*	685	***9.84***	**11.19**	**14.11**	**18.22**	12,483
Milimani	85.1	514	5.37	6.11	**7.7**	**9.96**	5,117
Kanyakwar	18.74	1014	1.18	1.35	1.7	2.19	2,222
Nyalenda	*188*	582	**11.87**	**13.58**	**17.01**	21.99	12,797
Manyatta	*223.1*	624	**14.12**	**16.05**	**20.22**	26.14	16,312
Wathorego	30.81	1035	1.94	2.21	2.7	3.5	3,618
Korondo	18.89	1754	1.19	1.36	1.7	2.21	3,874
Kogony	20.90	1480	1.32	1.5	1.89	2.44	3,618
Kasule	12.68	1871	0.27	0.30	0.38	0.49	925
Chiga	7.56	2083	0.15	0.18	0.22	0.29	614
Nyalunya	9.45	2035	0.2	0.23	0.28	0.37	750
Kodero	20.65	569	0.43	0.49	0.62	0.8	458
Got Nyabondo	17.77	843	0.37	0.43	0.54	0.61	599
Konya	19.73	1158	0.42	0.47	0.59	0.77	913

The numbers in **bold**[7] indicate the areas that have attained/surpassed threshold base flow density for sewerage
The numbers in ***bold and italic*** indicate the areas that are at verge of base flow threshold level
The numbers italic [8] indicate the areas that have attained requisite population density (p.e./ha) for sewerage

[5] Water and Sanitation Sector Investment Planning (SIP) for KIWASCO (2008).

[6] Figures are based on computation of population, water demand and waste water projections from LVSWSB (2008).

[7] The Kampala Sanitation Strategy and Master Plan threshold of 10 m³/day/ha base flow density for sewerage development is used in Kisumu city for comparison.

[8] The Ministry of Water and Irrigation Practice and Manual for Sewerage and Sanitation Services in Kenya use a population density threshold for sewerage development of 120 p.e./ha.

9.4 Sanitary Networks in Kampala and Kisumu

The four principal components of the physical sewerage networks in Kampala and Kisumu are sewage treatment works (STWs), gravity sewers, inverted siphons and pumping stations. Sanitary networks characteristics determine whether the sanitary systems can be called centralized or decentralized. The former option would inevitably require a number of both small and medium sized pumping stations designed with significant pumping heads in order to lift the sewage over the elevated boundaries of the natural catchment areas. Moreover, siphons will be required for flows to pass under major infrastructures.

All pumping stations (PS) in Kampala and Kisumu have two pumps each, intended to operate on a duty and stand-by basis, with an exception of Kibira road which has three. The pumps are operated manually because the automatic on/off switches are not working. Therefore, pumping stations are not operated at night due to security problems. As such, the night-flow is either by-passed to the environment, where a by-pass exists, or stored in the sewerage network, which often overflows through manholes when the system is full. Pump failures generally take a long time to be restored because purchases of spare parts, which are often imported, take up to a year. Moreover, electricity bills and/or fuel costs are huge. Most pumping stations, therefore, do not operate fully due to high costs, mechanical failure, high power tariffs and security concerns. Thus pumps often malfunction with downtimes ranging from 20% to 100%. The sewerage system in Kampala and Kisumu contains networks with the following characteristics (Table 9.5).

The sewerage network in Kampala comprises about 160 km of sewers with diameters ranging from DN 175–675 mm, with 64% of the pipes being less than DN 200 mm and 22% between DN 200–300 mm. In Kisumu, the total trunk sewer length for Central and Eastern Waste water District is 9.5 and 8.5 km respectively. The reticulation and trunk sewer network in Kisumu comprises pipes sized between DN 150–600 mm in central and DN 175–675 mm in eastern waste water districts. Sewerage schemes in Kampala and Kisumu cities are based on conventional sewers, which are typified by the laying of pipes along the road and designed for carrying large volumes of waste water for short periods. The sewers are generally dug at a reasonable depth with an average depth of about 2 m. However, in Kampala some 5% of the sewers are deeper than 4 m, with extreme depths of 15 m deep in Upper Kitante West sewer, making safe access for maintenance at such depths very difficult. Some 5% of the sewers are laid with a cover of less than 1 m in Kampala whereas in Kisumu some sections of eastern waste water district are open, where they use steel sewers. Manholes tended to be at about each 100 m and in some cases, distances between manholes are shorter making maintenance more easily. In Kampala, approximately 10% and 40% of the manhole covers are missing or buried respectively (NWSC 2004b; 2008) while in Kisumu all iron-based lids have been stolen (LVSWSB 2008). Lake Victoria South Water Service Board (LVSWSB) are replacing them with concrete slab covers.

The system of siphons is complex with each siphon consisting of two or three parallel pipes and in some cases duplicated and complicated junction chambers from

Table 9.5 Sewers and catchments in Kampala and Kisumu (NWSC 2004b; LVSWSB 2008)

Sub-catchment	Type of system	Pump head (m)	Average down-time (%)		Catchment area (ha)	(Total ha)
Kampala City			Pump 1	Pump 2		
High Level System						
High Level	Gravity + Siphon				350	
						1,265
Kitante West	Gravity + Siphon				250	
Kitante East	Gravity + Siphon				240	
Lugogo Valley	Gravity + Siphon				425	
Low Level System						
Low Level	Pumping	12.0	25	90	395	
						735
Bugolobi	Pumping	12.7	100	20	290	
East Bugolobi	Pump + Siphon	49.5	70	90	50	
Kisumu City						
Central District						
High Level	Gravity +Siphon					390
Low-Level System						
Sunset	Pumping	40	–	–		
Kendu Lane	Pumping	12	–	–		
Mumias Road	Pumping	10	–	–		
Eastern District						
High Level	Gravity					214

which siphons convey the sewage flow to the STW. It is extremely difficult to attain self-cleansing velocities with siphons. Siphons have a high tendency for blockage, are difficult to maintain with sewers having to be isolated and drained while blank flanges have to be removed before jetting. Moreover, every change of direction or obstruction is a potential blockage point (NWSC 2004b; LVSWSB 2005). The blockage problem is also exacerbated by sewer systems frequently receiving night soil and septic tank content discharges, leading to extreme solids concentrations in the network. Most of the siphons are located in built up areas and when blockages occur, sewage overflows along busy streets. Sanitary systems that rely on siphons and extensive pumping are difficult to maintain and operational problems will increase as population and economic growth pressures lead to increased sewage flows.

The high operation and maintenance costs and high chance of failure in extensive use of pumping stations and siphons has led to planning and development of sewerage based on catchments. This is the case with the development of an Eastern waste water catchment that drains to Nyalenda treatment ponds in Kisumu utilizing gravity sewers. Its development lead to abandonment of Martin's Dyke and Nairobi Road pumping stations and thus reduces operation and maintenance costs. The catchment approach in planning and development of sewage systems in for instance Nakivubo, Kampala, will lead to the replacement of all siphons by gravity sewers, whereas low level pumping stations will be abandoned altogether.

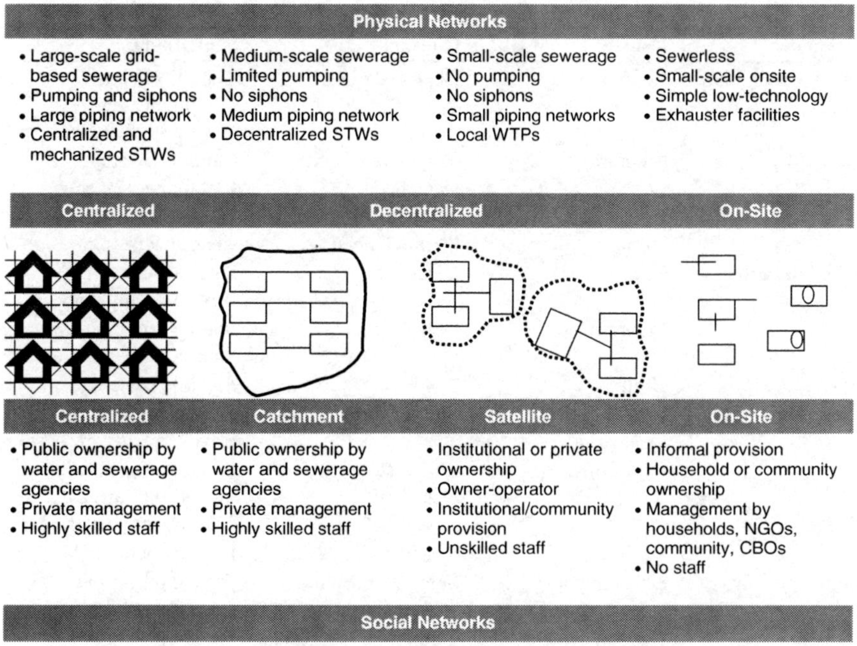

Fig. 9.2 Physical and social sanitary networks in Kampala and Kisumu city

Moreover, Kinawataka sewerage catchment development would lead to reduction of annual power costs from possible €50,000 to €7,000 per year while pumping station and construction costs would reduce from €2.29 million to €0.3 million (NWSC 2004b). Therefore, there are excessive constraints with extensive use of pumping stations and siphons, which is not in tandem with a high turn over of staff in sewerage profession. Therefore, catchment and satellite-based sewerage development is an attempt to remove or reduce significantly the use of pumping stations, eliminate siphons, save substantially on operation and maintenance costs, and make sewage systems easier to manage. Different networks show different social and physical network configurations (Fig. 9.2).

9.5 Sanitary Space Demands in Kampala and Kisumu

Demands for sanitary space include the position, location and land requirements for sewage systems in order to ensure accessibility to the facilities and compatibility to the existing land uses. It is difficult to implement one sanitation system in Kampala and Kisumu cities because of the nature of urban development that is in lack of formal urban development planning, adequate enforcement of planning and sanitation regulations, and of contiguous and homogenous development. Different

Table 9.6 Urban space structure and their sanitation systems in Kampala and Kisumu city

Development	Density	Housing	Sanitation type
Planned Urban Core	Low	Detached	Central sewerage Catchment sewerage Private septic tanks
	Medium	3–4 Storey flats Detached Semidetached	Central sewerage Catchment sewerage Private septic tanks
	High	Semidetached	Central sewerage
Planned Suburban	Low	Detached	Satellite sewerage Private septic tanks
	Medium	Detached	Shared septic tanks Satellite sewerage
	High	3–4 Flats	Satellite sewerage
Organic peri-urban	Low	Single storey	Private septic tanks Private quality pit latrines
	Medium	Single storey Detached	Private latrines Shared pit latrines Communal pit latrines
	High	3–4 Flats	Private septic tanks
Informal peri-urban	Densely packed	Semi-permanent housing	Private pit latrines Communal latrines
	Densely packed	Temporary structures	Private pit latrines Shared pit latrines Communal latrines Flying toilets

urban space structures have different sanitation types and consequently, sanitation is spatially defined by nature of urban development (Table 9.6).

The sewerage infrastructure development in Kampala and Kisumu is not based on any spatial threshold levels. Consequently, sewerage have been developed even on extremely low density areas such as 1–15 p.e./ha in Kololo in Kampala and in part of Milimani in Kisumu where density is 44.8 p.e./ha. Such neighbourhoods do rarely recover economic costs of sewer connection and thus are often subsidized by default through application of fixed connection fees instead of proportionate charges based on population or length covered. Such unsustainable sewerage can be curbed through establishment of density thresholds. The establishment of thresholds is to ensure environmental protection since at certain densities household on-site technologies reaches saturation limits while sewerage becomes cheaper.

Physiographic factors also influence the kind of sanitation technologies and networks to be adopted. Kampala city has grown from seven hills in 1900s to 25 presently in the metropolitan area while Kisumu grew from one drainage catchment in 1900s to three presently. Consequently, centralized sewerage becomes more difficult since it will entail costly transfer of flows from one drainage catchment to another. Soil and permeability conditions are also different within the city boundaries. For instance the soil thickness in Kampala ranges from 1–2 m at hilltops while at the flanks of the hills it increases steadily up to 5–7 m at the base of the slopes in the transition to the valleys.

Valley bottoms – although consisting of over 7 m deep soils – are characterized by high water tables and unstable soils making on-site sanitation based on percolation properties unsustainable because of permanent saturation and frequent collapse of latrines. Permeability of hill type soil is generally low with estimated values of 10^{-5} m/s (NWSC 2004b). Hillside soils are deep and thus pit latrines and septic tanks can be sustained. In Kisumu informal settlements of Nyalenda, Manyatta, Bandanai and Obunga, the average depth of wells ranges between 5 and 6 m with water levels rising to about 3 m deep and sometimes to ground surface in the Bandani and Obunga areas (LVSWSB 2005). If the water table is within 1 m of the ground surface, pit latrines, (unless they are connected to condominial or small bore sewerage systems) and reed odourless earth closet (ROEC's) are unfeasible because it poses environmental pollution and public health threats (MWI 2008). They may be feasible if the soil is sufficiently permeable such that the water level in a pit is not less than 0.5 m below the ground surface (World Bank 1983 in MWI 2008). However, even then, N leaching to groundwater remains an environmental and human health threat, meanwhile the resource is lost for reuse. Therefore, on-site sanitation is more problem mitigating than problem solving and often seen as transient. However, the transition from on-site to centralized sewerage is increasingly becoming a mirage with application of high planning and design standards as well as reliance on public provisioning systems. Thus there is need for a new approach to sanitation provisioning such as Modernized Mixtures, as discussed in the next paragraph.

Earlier land use plans in Kampala and Kisumu located sewerage works at the edge of the master plan area making the use of siphons inevitable in order to cross over major infrastructures such as the railway line in Kisumu and Kibira road in Kampala as well as use of extensive pumping stations to convey flows from various sub-catchments or lift to inlet works or areas where it can flow by gravity. However, developments have occurred beyond the master plan area and around the treatment works. The space use requirement for buffer zones to avoid nuisance constitutes a ring of preferably 500 m around waste water treatment works, away from any settlement, down wind from the community they serve, and away from any likely area for future expansion (Mara et al. 1992; LVSWSB 2005). Such buffer zones make waste water treatment works, as far as land use planning is concerned, repugnant to other urban land uses. As such an integrated approach is necessary with suitable sanitation treatment technologies, located upstream and operated close to settlements through flexible land use design that blends the neighbourhood, environment and STW sustainably.

9.6 Reconfiguring Sanitation Flows, Networks and Spaces in Kampala and Kisumu City

Our findings in Kampala and Kisumu show that there is need for reconfiguration of sanitary flows and networks against the background of existing spaces. Sanitation reconfiguration leads to embracing Modernized Mixtures elements to sanitation infrastructure development and management. Such mixtures entail a mix of scales, strategies, technologies, payment systems and decision-making structures that better fit the physical and human systems for which they are designed (Spaargaren et al. 2005). Modernized Mixtures involve a flexible mix of conventional and nonconventional approaches, planning and design, and institutional structures, which lead to shifts in provisioning strategies brought about by parallel development of public, private, institutional, community and household sanitation infrastructures or small-scale entrepreneurs (SMEs).

The first reconfiguration is a shift from centralization to decentralization both in terms of infrastructure development and management. The starting point of decentralization of public provisioning is a catchment approach to sewerage development and management. The drainage catchments (where in Kampala one catchment is composed of a number of hills) become the delineation boundary for decentralization and thus defining the scale of sewerage and location of treatment works. Another decentralization point is in suburban development, where settlements and sanitation infrastructures are developed by institutions and real estate companies that utilizes satellite sewerage, which is managed locally by the developers.

The second reconfiguration is a shift from exclusive adoption of conventional, high planning and design sewer standards to a mix of conventional and nonconventional sewerage. For instance in Kampala, hilltop soils are shallow and suitable for shallow sewers while hillsides with deep soils can easily be connected via conventional gravity sewers. In Kampala, where 17.5% of households are connected to septic tanks, and in Kisumu where most of the rich neighbourhoods of Milimani and Migosi are connected to septic tanks, settled small-bore sewers and a 50% tariff rate may lead to more cooperation, connection rates and sewerage coverage. The current approach of uniform sewerage charges in Kampala has led to refusal by residents connected to septic tanks to connect to the public sewer even within the mandatory sewer connection distance. The approach is typical conventional where a black–white choice with nothing in between is given. In areas with shallow rock, shallow sewers are more cost-effective than on-site systems (MWI 2008). In Kisumu, simplified condominial sewerage is envisaged for the slum upgrading programmes in Nyalenda slum and settled small-bore sewers where waterborne community sanitation blocks serve a high population (LVSWSB 2005).

The third reconfiguration is the use of base-flow and population density thresholds to determine areas and type of sanitation to adopt. Threshold levels, though contestable, provide a general framework for mapping out present and future areas for sewerage. For instance the Sanitation Strategy and Master Plan for Kampala (NWSC 2004a, b) and Practice Manual for Sewerage and Sanitation Services in Kenya (MWI 2008) use fixed population densities of 200 and 120 p.e./ha respectively, with potential base flow of at least 10 m^3/day, which is sufficient to sustain sewerage flows per hectare. However, Kampala Sanitation Program Feasibility Report applies variable saturation population densities where it is assumed that low, medium and high-income zones have a saturation of 50, 250 and 450 p.e./ha (NWSC 2008). Sinnatamby (1983) notes that 160 p.e./ha is the critical threshold for adoption of simplified sewerage instead of on-site sanitation. The rational behind population density is that all forms of piped networks demonstrate marked reductions in unit household costs as the density of settlement increases because the same length of pipe-work serves an increased number of houses. The population density at which this transition takes place varies with the physical conditions of the settlement, such as soil permeability and topography (Sinnatamby 1983; UNCHS 1986; MWI 2008). House on-site technologies maintain a constant unit cost irrespective of the density of settlement, thus at a given density of settlement, piped networks become more economical than on-site systems. For instance, in new property development, sewerage connections might be cheaper than septic tanks (NWSC 2004b; MWI 2008), whereas in the case of an existing property already served by a septic tank the discounted costs over 10 years resulting from by-passing the septic tank and connecting to a new sewer is much higher than if the property remains on a septic tank (NWSC 2004b). Alternatively, septic tank effluent connected to a small bore sewer system reduces costs significantly resulting in mitigating environmental problems on the long term. Areas with shallow rock, shallow sewers are more cost-effective than on-site systems at population densities as low as 110 p.e./ha (MWI 2008). The introduction of thresholds levels for sewerage development attempts to ensure that sewerage is cost effectively developed. Where population-density is extremely high, the area is sewered on public health and environmental protection grounds, thus giving a window of opportunity for poor neighbourhoods. Apparently, threshold levels are a suitable decision-making tool at strategic city level and where information is incomplete. Figure 9.3 provides an indication where each sanitation systems/technology is generally applicable.

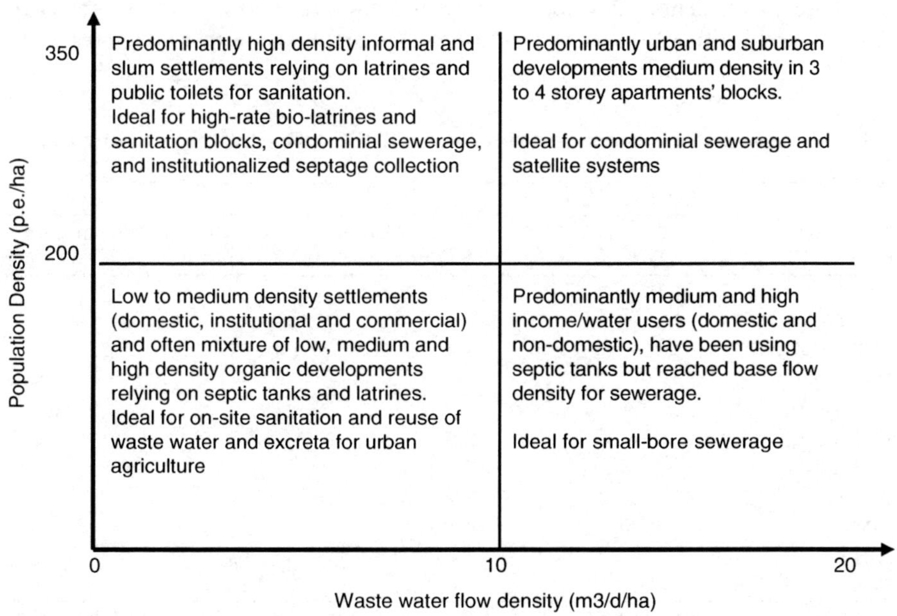

Fig. 9.3 Criteria for selecting areas to sewer (Modified from NWSC 2004b)

The fourth shift is from traditional approach to treatment works to a mixture of treatment process options. Traditionally in East African cities, sewerage modernization means a construction of a large but compact mechanized centralized STW. However, over the years, waste stabilization ponds (WSPs) have become more popular and widely applied. This shift from conventional to WSP translates into substantial land requirements which are not readily available. In addition, the required conveyance sewers can be very expensive. To arrive at suitable treatment and optimal land use requirements a flexible mix of process options is used. For instance, the proposed Nakivubo treatment process options result in different land use requirements (Table 9.7). The table only includes system alternatives that do not require mechanized aeration.

The land requirements given in Table 9.7 indicate that the overall land area requirements range from about 5–70 ha depending on treatment option and discharge standard. The first STW modernization includes conventional attached growth systems which were aimed at carbon removal. However, at present environmental regulations demand in addition pathogen and nutrient removals. To meet the new discharge demands, intelligent and flexible mixtures, which combine different technologies and which are meant to function independently in process options, are imperative as exemplified in Table 9.7. This not only reduces land use requirement substantially but also meets most of the discharge requirements at lower operation and maintenance costs. This mixture of concepts and technologies seems to provide an indication on the future direction in waste water treatment works development in East African cities. Moreover, upstream location of STWs challenges the 500 m buffer zone by sewerage systems in Kampala and Kisumu that borders residential developments with a less than 50 m wide buffer strip. Consequently, a flexible mix of land use design and location that integrates waste

Table 9.7 Land use requirements for the Nakivubo STW based on discharge standards (NWSC 2004b)

	Land size (ha)		
Technology process	Carbon	Carbon + faecal coliform	Carbon + faecal coliform + nutrient
PST-TF	4.3	–	21.2
PST-TF-M	4.3	19.4	–
PST-TF-M-W	4.3	19.4	29.2
UASB-F	22.0	–	–
UASB-F-M	22.0	27.3	–
UASB-F-M-W	22.0	27.3	44.2
UASB-F-W	17.0	22	33.9
A-F-M	39.8	39.8	–
F-M	68.9	68.9	–
A-F-W	–	–	39.8
F-W	–	–	49.0

Symbols: A (Anaerobic), F (Facultative), M (Maturation), PST (Primary Sedimentation Tanks), TF (Trickling Filter), UASB (Up-flow Anaerobic Sludge Bed), W (existing 16.9 ha natural wetland)

water reuse and valorization close to residential and urban agriculture areas needs to be embraced. The intelligent and flexible mix of technologies and land uses will allow, just like compact STWs that do exist in urban areas elsewhere, operation without odour nuisance less than 50 meters from neighbouring settlements. At present only compact combinations of UASB with trickling filters are exploited in countries like Brazil and Mexico very close to residential developments without odour nuisance. Therefore, one of the feasible compact technology mixtures for Nakivubo and other planned catchment treatment works is a combination of UASB-TF. Such combination reduces land use and could be operated at decentralized scale. Moreover, a more novel technology combination such as UASB + Biotower trickling filter would show higher levels of BOD removal, full nitrification, and significant N and pathogen removal.

9.7 Conclusions

This chapter has analyzed the development of sewerage infrastructures in East Africa by utilising a framework of flows, networks and spaces. Sanitary configuration in Kampala and Kisumu is brought about by different scales, technologies and institutional structures resulting in mixtures. There is a trend towards medium-large-scale public sewerage development in East Africa based on catchments. In 2030s in Kisumu and Kampala three and four waste water catchment districts (that have met the population and base flow densities) are envisaged respectively. However, there is also a parallel trend towards development of private and quasi-public satellite sewerage systems outside public sewerage areas. The development of three to four flats, and significant number of neighbourhoods on septic tanks, calls for adoption of unconventional sewerage that is simplified condominial and small-bore sewerage for flats and septic tank properties, respectively. Moreover, threshold values indicate where there is sufficient waste water generation and/or population density to sustain particular type of sanitation technology or system against the background of urban spatial structure and form in order to protect the environment and public health. The key to sustainable sanitation is to recognize and utilize the co-existence of different sanitation configurations rather than trying to integrate all configurations in a single format of infrastructure management. To do this, the approach of Modernized Mixtures as an interplay of flows, networks and spaces may help not only to analyse current sanitation management, but also to design new infrastructures and modes of governance for providing sanitation for all. However, the lack of uniform criteria and thresholds in the two cities calls for harmonization of strategies, standards and regulations applied in East African cities at large.

References

KIWASCO. (2008). *Water and Sanitation Sector Investment Planning (SIP) (2008)*. Kisumu: Kisumu Water Supply and Sewerage Company (KIWASCO).

LVSWSB. (2005). *Kisumu water supply and sanitation project: Feasibility report*. Kisumu: Lake Victoria South Water Services Board (LVSWSB).
LVSWSB. (2008). *Kisumu water supply and sanitation project: Sewerage design report*. Kisumu: Lake Victoria South Water Services Board (LVSWSB).
Mara, D. D., Alabaster, G. P., Pearson, H. W., & Mills, S. W. (1992). *Waste stabilization ponds: A design manual for Eastern Africa*. Leeds: Lagoon Technology International Ltd.
MWI. (2008). *Draft practice manual for sewerage and sanitation services in Kenya*. Nairobi: Ministry of Water and Irrigation (MWI).
Nilsson, D. (2006). A heritage of unsustainability? Reviewing the origin of the large-scale water and sanitation systems in Kampala, Uganda. *Environment and Urbanization, 18*(2), 369–385.
NWSC. (2004a). *Sanitation strategy and master plan for Kampala city: Volume 1 – Executive summary*. Kampala: National Water and Sewerage Corporation.
NWSC. (2004b). *Sanitation strategy and master plan for Kampala city: Volume 2 – Main report*. Kampala: National Water and Sewerage Corporation (NWSC).
NWSC. (2008). *Kampala sanitation program: Feasibility study project concept report*. Kampala: National Water and Sewerage Corporation.
Sinnatamby, G. S. (1983). *Low-cost sanitation systems for urban peripheral areas in Northeast Brazil*. Ph.D. thesis, University of Leeds, Leeds.
Spaargaren, G., Oosterveer, P., Buuren, J., & Mol, A. (2005). *Mixed Modernities: Towards a viable urban environmental infrastructure development in East Africa*. Wageningen: Environmental Policy Group, Wageningen University.
UNCHS. (1986). *The design of shallow sewer systems*. Nairobi: United Nations Centre for Human Settlement (UNCHS).
World Bank. (1983). *Water supply and sanitation project preparation handbook*. Washington, DC: World Bank.

Chapter 10
Meeting the Sanitation Challenge in Sub-Saharan Cities: Lessons Learnt from a Financial Perspective

Jérémie Toubkiss

Abstract The very low coverage rate makes sanitation one of the key issues at stake in Sub-Saharan Africa, particularly in overcrowded areas such as peri-urban zones and slums. Among the range of technical, institutional and social issues that arise from the sanitation crisis hitting this region, financing is of key concern, particularly in terms of how to build sustainable local financing mechanisms. ODA remains indispensable for financing major infrastructures that are too expensive for most African governments. Its scope is limited, however, when it comes to supporting on-site and semi-collective sanitation programmes that require long-term financing. This chapter discusses the context and the various financing mechanisms that have been tested in the field to encourage household investments. While microfinance is often restrictive and inaccessible to the poor, household subsidies are easier to manage and make installations more affordable, particularly for the poorest. Sanitation financing faces two main challenges. Firstly: how to set up financial mechanisms that effectively drive up household investments for on-site sanitation facilities? Secondly: how to finance operation and maintenance costs of waste water and sludge evacuation and treatment facilities? A sanitation surcharge on existing water services appears to be a sound example of a sustainable and effective local financial tool for sanitation.

10.1 Introduction

Whereas access to drinking water has been recognized as a major concern in the scientific community, the media and among decision makers for a long time, the sanitation sector has remained largely ignored. However, the situation is very alarming, especially in Sub-Saharan Africa where the coverage rate is the lowest and where MDG (Millennium Development Goal) 7, target 10, aiming at reducing the

J. Toubkiss (✉)
Hydroconseil, 198, Chemin d'Avignon 84470, Châteauneuf de Gadagne, France
e-mail: toubkiss@hydroconseil.com

B. van Vliet et al. (eds.), *Social Perspectives on the Sanitation Challenge*,
DOI 10.1007/978-90-481-3721-3_10, © Springer Science+Business Media B.V. 2010

number of people without sustainable access to sanitation by 50%, will not be reached by 2015. Sanitation is hereby defined as the management of household waste water and excreta – that is to say grey and black water. Different explanations are given for sanitation's poor progress in Africa, including technical, politico-institutional and cultural reasons. Yet one dimension remains little understood, namely the set of financing mechanisms applied to sanitation infrastructures at local level.

The following contribution focuses on this financial issue and thereby elaborates on the following questions:

- What is to be financed in Sub-Saharan Africa and how much will it cost?
- What are the challenges raised by the sector's financing needs?
- Which solutions are usually promoted and which ones are the most promising?

Most material presented in this chapter is drawn from existing literature and policy documents. However, answers to the last question, outlined in the final part of the chapter, will rely on the first lessons learnt from several in-depth case studies carried out by Hydroconseil in West Africa. They address the issues of private investment fostering, cost recovery and involvement of local authorities.

This chapter starts with a brief presentation of the sanitation crisis in Sub-Saharan Africa and the main obstacles complicating its solution. As mentioned above, one specific challenge is to finance sanitary infrastructures and their operation and maintenance. This question is discussed in the second section, followed by an insight into potential solutions, as suggested by key experts. The next section makes a review of some of the findings from Hydroconseil/pS-Eau's ongoing study in West-Africa. In the final section, financing prospects for environmental infrastructures are sketched.

10.2 The Sanitation Crisis in Sub-Saharan Africa

10.2.1 Africa Lags Behind

The coverage rate for improved excreta disposal (black water) in Sub-Saharan Africa is 38% compared to a 50% average in the developing world.[1] Though very low, this figure is probably overestimated because of the statistical method used by the WHO-UNICEF Joint Monitoring Programme and the general weakness of national statistics.

Save any radical change, the coverage rate is bound to keep on decreasing for years to come. As the demographic growth rate in Sub-Saharan Africa is very high (2.4%, that is to say 15 million additional inhabitants every year), the number of people without access to adequate sanitation will rise by 91 million between now and 2015. Reaching MDG Target 10 would therefore require an extra 35 million people per year to access sanitation (WHO 2008).

[1] JMP, 20th of March 2008: www.who.int/mediacentre/news/releases/2008/pr08/fr/index.html.

The coverage rate is higher in urban areas than in rural ones (53% against 28%). Yet the problem is actually more acute in urban conditions due to the combination of soaring demographic growth and surging urbanization. By 2015, the majority of the African population will be living in cities. Predictions are that between 1990 and 2015, rural population without access to sanitation will have been reduced by 25% whereas it will have increased by 50% in urban areas. To make it worse, it is even more troublesome to be deprived of latrines or to be unable to properly evacuate waste water in densely populated cities than in villages.

The situation in peri-urban areas and slums is even more problematic, because they are overcrowded and mostly constructed downstream of flood zones or of polluting sources such as residential and industrial zones. These areas seem to have been abandoned by national governments and municipalities while their inhabitants do not have the economic and political tools to claim their rights and express disapproval. For instance, many slum dwellers do not possess title papers that would enable them to prove their property right and demand local communities to provide them with infrastructure and services.

Hence, it seems that we are still a long way from substantially reducing the number of people without access to sanitation. According to the most optimistic forecast, Sub-Saharan Africa will reach MDG Target 10 only by 2108.[2]

10.2.2 Level of Service

Three "levels of service" exist in Africa. Large-scale collective sanitation facilities consist of public sewerage systems and currently can only be found in centres of large cities. At the moment, these centralized sewerage systems only service between 5% and 10% of the African population. They are very expensive to build and costly to maintain. Moreover the very low water consumption in African cities sharply curbs down sewers' performance. These difficulties, among many others, explain why the existing centralized networks are not well-maintained and/or run down (see also Letema et al. in Chapter 9 in this volume).

The intermediate level of service corresponds to semi-collective sanitation facilities like small bore sewers, which connect a limited number of plots. These models are growing popular in different parts of the world (Mara and Alabaster 2008). Although these systems are quite new in Africa, a number of them can be found both in West and East Sub-Saharan Africa (Mali, Senegal, Kenya, Uganda). Semi-collective facilities are often considered as a cheap alternative to sewerage, especially relevant in peri-urban areas. (See also Letema, Chapter 9 in this volume)

Finally, small-scale, on-site facilities such as pit latrines and septic tanks for individual households are the most widespread sanitation systems in Africa. These facilities remain the only type of sanitation for more than 90% of the African population both in rural and in urban areas.

---

[2] End Water Poverty Coalition, UN meeting on MDGs, 24th September 2008.

Considering that collective sewerage systems are costly, not widespread and that many hurdles hinder their implementation, one may infer that they are not a suitable solution to Africa's sanitation needs. The chapter concentrates therefore on how to increase access to on-site and semi-collective sanitation facilities.

10.2.3 Main Obstacles

Efforts to improve access to sanitation in urban African context are hampered by numerous obstacles. From a technical point of view, the main challenge is to find technologies that are appropriate to local conditions. In addition, those technologies must not be supplied by imports from the North, and shall remain affordable for poor governments, municipalities and populations. From an institutional point of view, roles and responsibilities for sanitation policies and their implementation are divided (or jointly shared) between ministries in charge of water, health, education, infrastructure or urban development, and often get muddled. In practice, therefore, frequent demarcation disputes or inaction can be observed. Lack of political and public concern with sanitation does not help either in getting this issue forward. Socioeconomic studies account for sanitation's low priority on governments' agenda by showing that populations seldom express a concrete demand for it.[3] This does not mean that sanitation is not a problem, but rather that most people have other priorities, such as economic survival, housing, health, education, or water supply. Finally, financing issues are also likely to restrain progress in access to improved sanitation in Sub-Saharan cities.

10.3 Financing Sanitation

10.3.1 What Is to Be Financed?

Financing sanitation is a complex issue, involving different activities, mechanisms and numerous different actors. To start the analysis, it is necessary to conceive sanitation as a three-stage process – collection, transport and treatment of waste water. Each of them is organized and financed in a completely different way. For each level of service (centralized, semi-collective and individual sanitation systems), these stages involve different activities, institutions and stakeholders and different technologies.

Furthermore, the process of developing, implementing and maintaining sanitary facilities includes a host of other categories of expenses which all have to be taken into account. Such categories include:

[3] See for instance various feasibility studies carried out by Hydroconseil for (water and) sanitation projects in West Africa (Burkina Faso, Senegal, Mali), and also Jenkins and Scott (2007) and a policy paper by the Government of South Africa (2002).

- Feasibility studies
- Investment/rehabilitation/renewal programs
- Operation and maintenance
- Training/capacity building of professionals and users
- Sanitation promotion and awareness campaigns
- Sector coordination

Solving sanitation problems in a sustainable manner requires the execution and financing of all different stages.

10.3.2 How Much Will It Cost?

The investments and activities necessary for implementing and managing adequate sanitation infrastructure in Africa require substantial financial resources. In the AMCOW/WSP/AfDB (2008) report, figures from Country Sanitation Reviews (based on national estimates, national investment programs or medium-term expenditure frameworks) have been compiled and extrapolated where data are missing in order to assess investment requirements. Results reported in this study suggest that approximately 26 billion USD are needed to achieve African national sanitation goals. This amount is comparable to other recent macro-level assessments which calculated that an approximate 23–50 billion USD would be necessary over the 2000–2015 period to reach the goal set in MDG Target 10 regarding sanitation (that is 1.5–3.4 billion USD per year depending on the estimates). This investment represents up to three times the amount required to reach the MDG set for access to drinking water.

However, these estimates only take into consideration the cost of collection and – sometimes – treatment of waste water, thus ignoring the cost of transport. Furthermore, they do not encompass all categories of expenses mentioned in the previous section. Most countries exclude the following items from investment needs assessments: feasibility studies, operation and maintenance, capacity building, hygiene education and sanitation promotion, policy formulation, planning, monitoring and regulation. Consequently, the full costs of reaching MDG Target 10 on sanitation will be considerably higher than the sums mentioned here (Toubkiss 2006).

10.3.3 Obstacles to Financing Sanitation

Financing the sanitary sector in Sub-Saharan Africa implies to tackle numerous difficulties. First of all, most governments lack the necessary resources to finance the investments and activities required. They are already in debt and, as aforementioned, they have other priorities, like health and education. Besides, in spite of the decentralization process taking place in most African countries, many local authorities still do not have the financial resources to develop sanitation systems. Water and sanitation infrastructures and services have become a local authority's affair but this transfer of responsibilities from the central government has not been accompanied by a corresponding transfer of financial means and technical skills. Thirdly, relatively few NGOs are active or specialized in the sanitation sector, although some exceptions exist[4]. They seem to remain more interested in water supply. Indeed the water supply sector has more visibility, a nobler image and generates more revenues. It is also spurred on by transnational companies' cheerleading. Lastly, the constraints evoked in the previous section remind that financing sanitation is a very thorny issue.

Statements made in the latter sections will be buttressed in the next part by key technical and financial stakeholders' analysis.

10.4 Donors' Perspective

10.4.1 Camdessus Panel (2001–2003)

On the occasion of the third World Water Forum, an international panel chaired by former Director General of the IMF (International Monetary Fund) Michel Camdessus published in 2003 a report on "Financing Water for All" (Winpenny 2003).[5] The aim of this panel was to find ways to raise more funds for the water and sanitation sector. Its main recommendations were:

- ODA (Official Development Assistance) to the water and sanitation sector should be increased and become more effective
- Long term, local currency, sub-sovereign lending should be developed. Funds should thus be transferred to local authorities rather than the central governments
- Local capital markets should be supported
- Private sector participation should be promoted
- Decentralization should be better organized, by transferring technical and financial means to the local level, training local government staff and fighting corruption

10.4.2 Gurria Task Force (2005–2006)

The Gurria Task Force chaired by former Minister of Finance of Mexico and current Secretary General of the OECD Angel Gurria published a report on the occasion of the fourth World Water Forum in 2006 (Van Hofwegen 2006). This publication

[4] CREPA (Centre Régional pour l'Eau Potable et l'Assainissement à faible coût), WaterAid or WASTE.

[5] The Financing Water for All report and its main recommendations are available at the following url: http://www.financingwaterforall.org/index.php?id=1107.

can be regarded as a follow up to the work initiated by the Camdessus panel. Apart from the usual considerations concerning ODA and financing tools proposed by international financial institutions and bilateral cooperation agencies, it tries for the first time to assess financing issues from the 'demand side', that is from the perspective of developing countries, bearing in mind the difficulties they face to access financial resources and turning them into tangible outputs. It highlights the necessity to push sanitation higher on the political agenda (for instance by including it in the Poverty Reduction Strategy Papers) and the importance of promoting demand for water and sanitation. Furthermore the report underlines that it is paramount to set an efficient tariff system that altogether reflects the real cost of water, allows for a sustainable cost recovery and is affordable for the poor.[6]

10.4.3 European Water Initiative (2003–…)

The European Water Initiative (EUWI) must be mentioned here too since its "Finance" and "Africa" Working Groups carry out research and experiments on the same topic. They added to the above-mentioned recommendations the need for better investment planning and budgeting needed at national and local government level.[7]

10.5 Where Do We Stand Today?

The Camdessus and Gurria panels as well as the European Water Initiative have had the beneficial effect of drawing international attention on Sub-Saharan Africa's pressing needs for funds to the sector. Yet the recommendations laid down are also too distant from real financing problems to offer concrete solutions. As a consequence, they are only partially relevant. A first reason why recommendations tend to miss the point is that the reports pay more attention to the water sector than to sanitation issues. Moreover, their recommendations are more appropriate to the larger developing countries that are able to access ODA lending and have a significant financial market. In Sub-Saharan Africa this concerns only a few countries, basically South Africa, Nigeria, Ivory Coast and Kenya. These reports mainly focus on big infrastructure projects, like collective sanitation and waste water treatment plants, although they represent a very small minority of existing – and necessary – equipments. The real challenge – that consists in financing on-site (and semi-collective) sanitation – is hardly addressed. Broadly speaking, the literature generally overlooks the question of how to finance poor urban population's access to on-site and semi-collective sanitation facilities. Finally, these expert groups

[6] Available at www.financingwaterforall.org.

[7] See Finance Working Group, www.euwi.net and the Africa Working Group, www.euwi.net.

search primarily for means to increase and ease international financial flows towards the water and sanitation sector. Yet, one may wonder whether financial resources' scarcity is the real problem. Mobilizing international funds for major investment projects is not a major problem. Indeed, good projects always find backers and there is even a sort of competition between the various multilateral financial institutions, as well as with bilateral cooperation agencies and commercial banks. Some institutions are even complaining about it, thereby forgetting that this is a mark of abundance and works in the population's interest. Furthermore, one could also mention examples of weak, poorly-designed or ill-adapted projects that have easily found funding.

In reality, the major problem with regards to on-site sanitation facilities is to set up financial tools (micro-finance, reimbursement facilities, household subsidies etc.) that prompt private investment effectively and sustainably. Concerning bigger infrastructure, it proves most challenging to finance operation and maintenance costs whereas the issue gets little attention. Yet, anticipating recurrent costs is key to the sustainability of (semi-)collective sanitation infrastructures.

10.6 pS-Eau and Hydroconseil Financing Sanitation Case Studies

Conscious of the problems linked to on-site sanitation facilities development and maintenance, the French Ministry of Foreign Affairs decided to fund a study addressing these two issues. This study has been coordinated by the French NGO pS-Eau and carried out by the consulting company Hydroconseil. This section briefly introduces this research and reports on its first findings.[8]

10.6.1 Objective and Methodology

The objective of the large-scale pS-Eau/Hydroconseil field study is to gather quantitative and qualitative information about the financing of sanitation programmes in Sub-Saharan cities. Emphasize was put on access to on-site and semi-collective sanitation infrastructures, and on maintenance of these infrastructures in urban and peri-urban areas. The results from this research will allow to draw lessons on factors of success and failure.

The study started in 2007 with a synthesis of the existing literature and was supplemented by 12 case studies reflecting the diversity of financing mechanisms and sanitation technologies (Table 10.1). With the help of a common template, these case studies were compared and lessons were drawn to produce a decision-making

[8] For more details on case studies and available papers linked to this study: www.pseau.org.

Table 10.1 List of case studies

Country	City	Target area	Case study's focus
Mali	Bamako	Hippodrome	Small bore sewer
Mali	Bamako	Banconi	Small bore sewer
Mali	Bamako	Citywide	Faecal sludge management and treatment by private operator "GIE Diabeso Saniya"
Mali	Bamako	Commune IV	Faecal sludge treatment plant Samanko II
Mali	Bamako	Commune II	Faecal sludge treatment plant Industrial Zone
Burkina Faso	Bobo-Dioulasso	Citywide	Strategic Sanitation Plan (PSAB)
Burkina Faso	Ouagadougou	Dapoya	Partnership ONG ENDA-rup/MFI FCPB on micro-financing on-site sanitation
Senegal	Rufisque	Castor, Arafat, Diokoul	ENDA-rup on-site and semi-collective sanitation project
Senegal	Dakar	Peri-urban areas	On-Site Sanitation Program in Dakar's Peri-Urban Areas (PAQPUD)
Niger	Filingué	Citywide	On-site sanitation and faecal sludge evacuation
Niger	Dogondoutchi	Citywide	On-site sanitation and faecal sludge evacuation
Uganda	Kampala	Citywide	On-site sanitation project and sludge management reform

support guide for local, national and international stakeholders (municipalities, NGOs, decentralized cooperation, central governments and donor agencies). These cases cover the three stages of the sanitation process (collection, evacuation, treatment) and include every related expense involved in sanitation activities.

10.6.2 Preliminary Findings

Different conclusions may be drawn from the case studies, but an overriding impression which emerges from the work so far is that the main problem in financing sanitation infrastructure is to be found at the local level. It does not so much stem from the need to increase funds supply (in quantity), but rather lies in the need to improve demand (in quality). Therefore the priority should not be to seek further financing among donors, but to better design the projects beforehand. It is indeed locally that the challenge of financing sanitation must be taken on.

10.6.2.1 Stimulating Private Investment: Microfinance Versus Household Subsidies

To encourage the development of household facilities, the main challenge is to find effective means of supporting private investment. *Micro-finance* is often promoted by development partners but raises more problems than it solves. The experiences of ENDA-rup in Rufisque and Ouagadougou and the case of payment facilities set up for mini-sewer users in Bamako make this clear. Micro-finance schemes always include interest charges which (even if subsidized) necessarily increase the cost-price of the equipment used by households, instead of making it more affordable. Furthermore, poor households' incomes are not sufficiently high or regular to ensure proper reimbursements. Neither does the sanitation facility obtained generate revenues that help households to make the reimbursements. Micro-credits for on-site sanitation facilities should therefore rather be considered as a consumer loan than as an investment credit. In fact, promoting this strategy increases the risk of putting poor households into debt. A practical problem is to find the balance between low monthly reimbursements spread over long periods compared with a shorter period of higher payments. The first option seems attractive but it is doubtful whether households will continue to make payments for such a long time, while the second option would probably imply excessive monthly charges for most households. Another problem is that eligibility criteria for micro-credit are often restrictive. The necessary conditions, such as filling out a form, building savings beforehand, and finding guarantees, cannot be met by the poorest households. Finally, handing out credits requires additional monitoring as there is no guarantee that money will in fact be used to invest in sanitation.

Alternatively, the examples of the peri-urban sanitation improvement programme in Dakar, and the strategic sanitation plan in Ouagadougou and Bobo-Dioulasso show that *household subsidies* are easier to manage and make the equipment more affordable, particularly to the poorest households (Saywell & Fonseca 2006). Subsidies generally take the form of equipment (i.e. a slab) provided to the household for free. The first advantage of household subsidies is obviously that they improve households' capacity and willingness to pay since they decrease equipment's final cost. Furthermore, they are easy to manage where a good supply chain exists (for slabs, siphons, etc.) and they avoid cash flows – and the related risk of misappropriation. In Ouagadougou and Bobo-Dioulasso, around 900,000 people have been provided with sanitation facilities in the last 15 years using a limited,[9] targeted and carefully designed household subsidy scheme.

Recently, an *anti-subsidy* trend has been developing, relying on several NGOs and donors' experience in South-East Asia, where they promoted sanitation facilities without any household subsidy. These 'open-defecation free community' and 'community-led total sanitation' approaches try to enhance the feeling of shame – which is supposedly linked to the practice of open-air defecation – and create a community-based pressure (for example in the form of a competition between neighbouring villages) as incentives for individual households to invest in their own facilities (WSP 2007). WaterAid and UNICEF intend to adapt these methods to the

[9] Between 10–35% of the capital cost depending on the type of facility.

context of East and West Africa, while DANIDA (Danish International Development Agency) and the European Union have expressed some interest. This innovative strategy seems interesting as financing household subsidies requires large amounts of funds. Donors are thus looking for such cost-efficient ways to promote sanitation. The difficulty is that most households in West Africa already have some sort of traditional latrine at their disposal, and sanitation equipments are more costly than in East Africa and South-East Asia, so that their willingness and capacity to pay for improved facilities is probably much lower. However, it may be too soon to draw a final conclusion on advantages and drawbacks linked to these approaches and more time is needed to get feed-back from large-scale field experiences.

10.6.3 Cost Recovery

In order to recover the cost of the facility and finance its maintenance, which is a condition for the sustainability of collective or semi-collective infrastructure (mini sewers, waste water and sludge treatment plants), various strategies are possible:

10.6.3.1 Selling Compost from Recycled Faecal Matter from Sludge Treatment Plants

Co-composting and recycling are often failures (as shown by the cases of Bamako and Dakar) because potential users of compost are reticent to purchase and use faecal matter. Moreover, in urban areas the market for compost is often insufficient to mop up supply. Furthermore, it is often necessary to combine faecal matter with organic waste to produce compost. This implies to organize a complete supply chain (from input collection to output distribution). In the end, the compost generally becomes more expensive than conventional industrial fertilizers or soil improvers. Lastly, the use of sludge in agriculture is not well mastered and may involve risks for human health (see also Grendelman and Huibers, Chapter 12 in this volume).

Some towns have experimented with introducing a fee charged on vacuum trucks operators to use the faecal sludge treatment plant and that way cover the plant's operational costs. The fee is charged on "vacuum" trucks' load (a certain amount per cubic meter). Making this strategy successful requires building large plants – large enough to receive sludge from the whole town – and more accessible than the unauthorized dumping areas. It is also recommended to penalize unauthorized dumping strictly in order to encourage vacuum trucks to dump their load at the plant. The cases of Dakar or Dar es Salaam suggest that this approach is promising. In those cities the volume of sludge transported to treatment plants has grown significantly and the revenue generated by the fee covers most if not all operation and maintenance costs (WUP 2001; see also Béréziat 2009). The Office National de l'Eau et de l'Assainissement of Burkina Faso intends to test this model in two faecal sludge treatment plants in Ouagadougou.

10.6.3.2 Charge Households a Monthly Fee to Maintain Mini-Sewers and Their Small-Scale Waste Water Plants

Charging households a monthly fee to finance the maintenance of semi-collective systems like mini-sewers and their treatment plants is also difficult as the experiences in Dakar, Rufisque and Bamako highlight. Indeed, in general, poor households do not set aside the money that would enable them to pay such a fee each month. This is particularly true if the fee is due as a payment in advance to finance repairs that are not necessarily needed immediately, as this generally is. Households are reluctant to save money for such fees because they have other priorities, like food and healthcare.

In addition, users do not feel particularly concerned with the maintenance of structures that are far away from their home (main drains, treatment plants etc.). All they want is to have their waste water evacuated from their premises. They do not want to worry about where it goes and how it is dealt with. Moreover, as public authorities or NGOs are usually at the origin of projects (rather than households themselves), it seems perfectly normal to users that these institutions should also pay for maintenance – just as the State has always paid for the maintenance of public infrastructures such as roads and bridges.

Therefore, we could imagine that households be willing to pay for infrastructure maintenance when this directly affects themselves (e.g. maintenance of the drains in front of their homes). On an occasional basis, that is when the need for maintenance is directly noticeable, they could gather the necessary funds by pooling resources together with their neighbours. This is how the system has gradually and naturally evolved in Bamako. In this model, the municipality should remain responsible for financing the maintenance of infrastructures that benefit the community as a whole or those aiming at protecting the environment (main network, treatment plants).

From these examples, we can infer a series of conclusions. The very first one is that financing operation and maintenance of sanitation infrastructure is tricky. We also noticed that the issue can be dealt with in different ways, using various tools. Lastly, problems stemming from sanitation financing vary whether it concerns on-site or semi-collective facilities (pS-Eau & Hydroconseil 2007).

10.6.4 Involving Local Authorities

Although the decentralization process in the water and sanitation sector is supposed to give local authorities a dominant role, they are almost absent in the projects and programmes studied so far. There is a need to involve them much more, especially in towns and rural areas where no other stakeholder has a clear strategy or concrete actions. But this remark is also relevant in urban areas, where local authorities should take up their responsibility for the maintenance of waste water and faecal sludge treatment plants and of semi-collective sanitation infrastructures. Their role could also be to make sites available for those facilities, which is not an easy task in densely populated neighbourhoods.

10.7 Conclusions

Household subsidies, rather than micro-credits, seem to be the most effective way forward in funding the improvement of on-site sanitation systems at scale, as argued in this chapter. Yet securing the necessary financial resources for such programs remains a challenge for most developing countries.

ODA remains indispensable for financing major infrastructures that are too large and too expensive for their costs to be covered by most African central and local governments. Yet, the ODA financing tools that are appropriate to finance large infrastructural projects like sewers and treatment plants are not well-adapted to fund on-site sanitation facilities. These equipments represent a scattered, difficult to plan, private investment, spread over a long period of time. Promoting it in the framework of large-scale programmes requires long-term, locally-available funding. In that regard ODA is an inadequate resource. In such instances the negative impacts of ODA become visible as it may create a large degree of dependency on external sources. For example: the peri-urban sanitation improvement programme in Senegal, which intended to provide more than 500,000 peri-urban residents with better sanitation facilities, had to be interrupted suddenly after 3 years when the operator realized that the World Bank's loan had run out. Today, 56% of the household requests that were initially registered are not fulfilled yet.

Such problems may be prevented through the introduction of a 'sanitation surcharge'. Several African countries have already implemented such a system (a.o. Burkina Faso, Senegal, Tunisia). A sanitation surcharge enables the water utility company to take a certain percentage from water users' bill finance sanitation activities. The advantage of this tool is that funding is internal to the water and sanitation system, continuous, predictable and most likely growing from year to year. In Burkina Faso, for instance, the income created this way is used to finance the Strategic Sanitation Plans in Ouagadougou and Bobo-Dioulasso (PSAO and PSAB), which has allowed to provide access to sanitation to almost 1 million people over the past 14 years. These cities have achieved by using local financing – almost without assistance or dependence on international donors.

PAQPUD and PSAO/B are examples of large-scale sanitation programmes. Such programmes are very rare in Sub-Saharan Africa. On the contrary, numerous small or micro-projects can be found in the field, with very little impact. Yet, enough experience has been gained to take the next step and scale up these experiments. There is certainly a need for further evaluation and capitalization, but it is time to step out of the 'pilot project' approach and improve urban sanitation facilities for larger numbers, if MDG 7, Target 10 is to be achieved in the foreseeable future.

References

AMCOW/WSP/AfDB (Water and Sanitation Programme/African Development Bank). (2008). *Can Africa afford to miss the sanitation MDG target? A review of the sanitation and hygiene status in 32 countries*. Washington, DC: World Bank – WSP.

Béréziat, E. (2009). *Partnerships involving small-scale providers for the provision of sanitation services: Case studies in Dakar and Dar-Es-Salaam.* UNESCO-IHE (with BPD and Hydroconseil).
Government of South Africa. (2002). *Framework for a national sanitation strategy.* Pretoria: Government of South Africa.
Jenkins, M. & Scott, B. (2007). Behavioral indicators of household decision-making and demand for sanitation and potential gains from social marketing in Ghana. *Social Science & Medicine, 64*(12), 2427–2442.
Mara, D. & Alabaster, G. (2008). A new paradigm for low-cost urban water supplies and sanitation in developing countries. *Water Policy, 10,* 119–129.
pS-Eau & Hydroconseil. (2007). *Case study: PAQPUD programme in Dakar* (unedited). Paris: pS-Eau & Hydroconseil.
Saywell, D., & Fonseca, C. (2006). *Microfinance for sanitation.* Well Factsheet. Retrieved June 3, 2009, from http://www.lboro.ac.uk/well/resources/fact-sheets/fact-sheets-htm/mcfs.htm
Toubkiss, J. (2006). *Costing MDG Target 10 on water supply and sanitation: Comparative analysis, obstacles and recommendations.* Montreal: World Water Council.
Van Hofwegen, P. (2006). *Task force on financing water for all; enhancing access to finance for local governments financing water for agriculture.* Marseille: World Water Council.
WHO (World Health Organization and United Nations Children's Fund Joint Monitoring Programme for Water Supply and Sanitation [JMP]). (2008). *Progress on drinking water and sanitation: Special focus on sanitation.* New York: UNICEF; Geneva: WHO.
Winpenny, J. (2003). *Report of the world panel on financing water infrastructure; financing water for all.* Marseille: World Water Council.
WSP (2007). *Community-led total sanitation in rural areas: An approach that works.* New Delhi: WSP.
WUP (2001). *Case study: Cesspool emptiers in Dar Es-Salaam.* WUP No 5. Abidjan: Water Utility Partnership for Capacity Building.

Part III
Perspectives from Farmers and End-Users

Chapter 11
Role of Farmers in Improving the Sustainability of Sanitation Systems

Håkan Jönsson, Pernilla Tidåker and Anna Richert Stintzing

Abstract The nutrient flow with excreta is one of the major plant nutrient flows in society. If these nutrients are recycled to arable soil in a safe and resource efficient way, the sustainability of the sanitation system can be increased. To achieve this, farmers using the excreta nutrients as fertilizers are just as essential as the toilet users. To get these recycling sanitation systems to function well, experience shows that it is very essential that the farmers participate already in the initial planning of the recycling sanitation system. Furthermore, the recycling sanitation system has many stakeholders and the recognition of the drivers and restrictions of each stakeholder is important for the functioning of the system as a whole. One important driver for many farmers is whether they can increase their business by taking responsibility for the collection and handling of the sanitation system fertilizer. The system has many stakeholders and it is for its long term performance, important that an arena for feed-back and interaction is organized.

11.1 Introduction

The nutrient flow with excreta is even today one of the major plant nutrient flows in society and was much more so before the introduction of chemical fertilizers. Human excreta were used as fertilizers in many countries before

H. Jönsson (✉)
Department of Energy and Technology, Swedish University of Agricultural Sciences (SLU), P.O. Box 7032, SE-750 07 Uppsala, Sweden
EcoSanRes, Stockholm Environment Institute (SEI), Kräftriket 2b, SE-106 91 Stockholm, Sweden
e-mail: hakan.jonsson@et.slu.se

P. Tidåker
Svenskt Sigill, SE-105 33, Stockholm
e-mail: pernilla.tidaker@sigill.lrf.se

A.R. Stintzing
Richert Miljökompetens, Åsögatan 140, SE-116 24 Stockholm
e-mail: anna@richert.se

B. van Vliet et al. (eds.), *Social Perspectives on the Sanitation Challenge*, 179
DOI 10.1007/978-90-481-3721-3_11, © Springer Science+Business Media B.V. 2010

chemical fertilizers were introduced. The powdery poudrette, produced from urine and faeces collected in cesspools, was a popular fertilizer around e.g. Paris and Gothenburg (Barles 2007; Wetterberg and Axelsson 1995). It has been estimated that as much as 40% of the nitrogen flowing into Paris in 1913 was recycled back to agriculture as fertilizer (Barles 2007). The loss of the valuable excreta fertilizers was in many places one of the major arguments against the introduction of the water flushed toilet. This argument lost most of its power with the introduction of the chemical fertilizers. The present linear flow of plant nutrients in our society, from arable soil through food to human excreta and then mainly to recipients water, atmosphere or landfill, is thus less than 100 years old in Europe.

At present only a very small proportion of the plant nutrients in sewage is recycled. This is partly due to the large scepticism present in many countries against the use of sewage sludge in agriculture (Aubain et al. 2002). The scepticism from the agricultural sector is not surprising, as the conventional system is not built with the preferences of the farmer in mind, and the farmer is not included as a stakeholder within the conventional sanitation system.

On the other hand, if waste water nutrients would be recycled to arable soil in a safe and resource efficient way, then the sustainability of the sanitation system and of society as a whole could be increased. Most of the waste water nutrients, but only a small fraction of the volume flow, are contributed by faeces and especially urine. Source separating and urine or black water recycling systems can therefore be efficient and increase the sustainability of a sanitation system and of society as a whole (e.g. Jönsson 2002; Tidåker et al. 2006; Bengtsson et al. 1997). These new systems are based on new toilets, that is urine diversion or vacuum toilets, where the urine or black water is source separated with as little dilution as possible. From the toilet the urine, or black water, is led through dedicated pipes to collection tanks for later treatment, transportation and then application as fertilizer on arable land.

However, yet important as these new technical installations are, new stakeholders need to be included in the sanitation system right from the start of the planning. The farmer is a key stakeholder as there will be no reuse without him, and thus the objective of the system will not be achieved. Additional new stakeholders are also the stakeholders in charge of the collection, storage and treatment of the recycled product and the stakeholders in charge of the distribution and spreading of the recycled product on the farmland. Frequently the farmer can take up several of these stakeholder roles, in addition to making the land available for the recycled fertilizer. He often has the equipment to spread, and often also equipment to collect, store and treat the product. This multi-tasking of the farmer is usually to the advantage of total system performance.

New stakeholders in these new sanitation systems are the farmers, the technical and environmental departments at the municipality and sometimes entrepreneurs carrying out tasks of the technical departments. The purpose of this paper is to share some Swedish experiences on how to best include farmers in these new recycling sanitation systems.

11.2 Methods of Analysis

This book chapter is mainly based on the results from a study of seven sanitation systems which recycle plant nutrients in Sweden. Three of the systems recycled source separated urine; two systems recycled black-water (i.e. toilet waste water), one system recycled a mixture of source separated urine and black water and one system recycled sewage sludge. The systems were explored by means of semi-structured interviews of one farmer (recycler) and of the system coordinator for each system. The system coordinator was often a municipal official. The study is described in full in Tidåker (2007). The results from this study have been supplemented by experiences and results from other projects. Our accumulated knowledge is presented in this chapter as motivated recommendations.

11.3 Investigated Systems

11.3.1 Case 1. Stockholm-Centralized System for Urine Spreading

Starting with a R&D project in 1996, the Stockholm Vatten company, responsible for water and waste water in Stockholm and several neighbouring municipalities, has had a centralized system using source separated human urine as a fertilizer on their agricultural land at Bornsjön, south of Stockholm. Several housing districts in the Stockholm region transported their source separated urine, at their own cost, to the three storage tanks, together 450 m^3, at Bornsjön. When the tanks were full, every second year the urine was spread as fertilizer for a crop, usually oats or barley. The Stockholm Vatten company paid for the spreading and analysed the nutrient content of the urine. Together the coordinator at Stockholm Vatten and the tenant farmer decided on a fertilization strategy, which enabled the urine nutrients to fully replace chemical fertilizers.

11.3.2 Case 2. Håga-Ecovillage with Neighbouring Farm

In 1999, the Håga Ecovillage, with 22 households using urine separating toilets, was finalized in Uppsala. An important motivation for the urine separation system was the use of the plant nutrients for cultivation of crops. Contact was early established with the cereal production farmer having a field close to the Ecovillage. The farmer, the only suitable close by, accepted using the urine, but only when paid for the cost of spreading it. The farmer preferred autumn spreading before ploughing to reduce soil compaction, although this lowered the utilization of nitrogen by the crop. The urine only sufficed to fertilize 1–2 ha which was a marginal fraction of the farmer's arable land. The farmer therefore did not find it worth while to modify his fertilization plan. The phosphorus was however available for coming crops.

One member of the ecovillage was responsible for all contacts with the farmer, and the municipality was not at all involved. As a result of frequent malfunctioning of toilets and dissatisfaction about sub-optimal nitrogen utilization from the side of the tenant-owner association, most of the households changed to conventional flush toilets in 2006–2007.

11.3.3 Case 3. Norrköping – Urine Separation with Strict Division of Responsibility

Substituting chemical fertilizer and decreasing eutrophying emissions were the two main reasons for the municipality to advocate urine separation as the main option for new on-site systems. In 2006, about 130 household had urine separation systems. Early contacts were made with a farmer, who stored the urine and reused it in a crop in the spring, without any payment either way. The recycling system was organized with a strict division of responsibilities, but without any single coordinator being responsible for the whole system. The environmental department was responsible for contacts with the households, the technical department for emptying the urine collection tanks in the autumn and the farmer for the storage and spreading. No analysis was made of the plant nutrients in the urine, and the farmer was therefore unaware about how much chemical fertilizer it could replace, but he did reduce the application of chemical fertilizers.

11.3.4 Case 4. Tanum – A Municipal Pioneer on Urine Separation

In 1998, Tanum started to require urine separation for all new houses with on-site sanitation systems, and in 2006 there were more than 400 permits for urine separation systems. The main motivations for urine separation were to decrease eutrophying emissions and to replace chemical fertilizer. The municipality established contact with farmers at an early stage. In 2006, seven farmers were engaged as entrepreneurs for collection, storage and spreading. For emptying the urine tank, the household could contact any of these farmers or the municipality. Contact with the closest farmer was recommended. The farmers charged the household for emptying the urine, and if the households contacted the municipality instead, a somewhat higher rate was charged, steering the household owners towards direct contact with the farmers. The farmers were free to set their fee and thus the fee differed between the farmers. Some of the farmers also collected and used black water from different households as fertilizer. Annually, the farmers were required to send documentation on which urine installations they had emptied. The municipal coordinator organized annually a follow up meeting with the farmers and the municipal stakeholders.

11.3.5 Case 5. Lund-Black Water Irrigation

Normally black water collected by the Lund municipality was treated in the municipal waste water plant. The aim of this project, which started in 2003, was to investigate how the black water, collected from a water protection area, instead could be used as a fertilizer for crop production. The black water was very diluted compared to source separated urine and to manure. Thus, the strategy was to spread it with irrigation equipment to energy crops. This made it possible to cover the plant nutrient requirement of the crop with the black water, and thus no mineral fertilizer was used. The municipality paid the farmer for using the storage facility and the irrigation equipment.

11.3.6 Case 6. Kvicksund-Liquid Composting at Farm Level

As part of its environmental profiling, vacuum toilets were installed in a school in Kvicksund, built 1998. The black water and the kitchen waste collected from the school were treated in a liquid composting plant on a nearby farm. The municipality was responsible for transport to the liquid composting plant, for handling major interruptions and economic commitments and for analyzing the plant nutrient content of the compost. The farmer was responsible for running the compost plant, storing and spreading of the compost. The compost was spread in the autumn, before sowing winter wheat. The reason for spreading in the autumn was that it minimized the risk for soil compaction. The farmer considered that this outweighed the fact that, with this fertilization strategy, only a small fraction of the nitrogen in the compost did replace chemical fertilizers.

11.3.7 Case 7. Sewage Sludge Recycling

In the Stockholm area, several waste water treatment plants have handed over the responsibility for handling, storing and sampling and finally agricultural reuse of their sewage sludge to the company Ragn-Sells. The company has agricultural experts in their organization and has long established relationships with a number of farmers. During 2006–2007, some 10–12 farmers in the region have received sewage sludge. The farmer interviewed had frequently received sludge for more than 20 years. He appreciated the nitrogen and phosphorus fertilizer value of the sludge and considered it economically profitable. He received the sludge, delivered and spread it on his field for free. The sludge was spread in the autumn, and thus the nitrogen was not fully utilized. As many food and fodder companies do not accept crops fertilized with sewage sludge, he had to choose which crop to fertilise with great care.

11.4 Analyzing Drivers and Restrictions for Stakeholders

The study revealed that the main drivers for nutrient recycling differed considerably between the farmers on the one hand and the project coordinators on the other. Most of the source separating recycling systems were initiated in the 1990s. Many of the source separation systems were motivated by the aim to close the loop of plant nutrients, and thus save on the finite resources needed to produce chemical fertilizers. In several cases, though none included in the study, the aim was to supply ecological farms with badly needed easily available nitrogen. Source separated urine was an allowed fertilizer in ecological farming in Sweden before EU membership. Thus, it was not surprising that most of the coordinators considered the replacement of chemical fertilizers a cornerstone of the system.

However, this high priority for replacement of chemical fertilizers was not shared by the farmers. It was only in the cases when the coordinator was in control of the spreading (the Stockholm, Lund and sludge cases) that the product was analyzed and thus the full potential of replacing chemical fertilizers could be realized. Nevertheless the nitrogen was not fully utilized in the sludge case, since the sludge was spread in the autumn. In the other cases, the coordinator only assumed that the recycled fertilizer replaced chemical fertilizers, but they did not analyze the content of plant nutrients in the product. In fact none of the farmers abandoned chemical fertilizers. This is in line with common Swedish farm practice, where only a small proportion of the farmers analyze the nutrient content of their manure. Furthermore, the nutrient flows with the recycled urine and black water were in all cases very small compared to the total nutrient flows on the farm. In most cases the flow only sufficed for fertilizing 1–2 ha per year, even though it sufficed for 5–10 ha per year in the Stockholm case. Thus, the main effort of the farmers was to streamline the handling of the urine and the black water, rather than to optimize the handling for maximum nutrient efficiency. This meant that the nitrogen was fully utilized only in the Stockholm and Lund cases. In the other cases, the spreading was done in the autumn, when there was no crop ready to take up the nitrogen. Thus, probably most of it was lost during winter. Phosphorus binds to the soil and stays fairly available for several years. This means that in most of the cases, the applied phosphorus had the potential to replace chemical fertilizer phosphorus. The exceptions were the cases, where the farms already were well supplied with phosphorus from animal manure, as was the case in Norrköping and Kvicksund.

When source diverting systems become more mainstream and the nutrient flows become larger, the economic incentive for the farmers will be much larger, to utilize the full nutrient potential of the fertilizers. A conclusion from the study was that when it is considered important to realize the full potential of replacing chemical fertilizers, then the merits of this needs to be discussed with the farmer. He should also be supplied with information on the nutrient content of the recycled fertilizers. Another conclusion was that the communication with the farmers functioned much better in cases where the coordinator knew farming practices and understood the conditions for the farmers well.

Most of the farmers considered the decreased eutrophying emissions a main advantage of the recycling sanitation system, which for source separating systems (urine and black water separation) is well supported with results from environmental systems analyses (Bengtsson et al. 1997; Jönsson 2002; Jönsson et al. 2005; Tidåker et al. 2006; Tidåker et al. 2007a, b, c).

One important restriction experienced by the farmers was the limited acceptance of the companies buying the farm produce. Many of these companies are in their policy statements very positive to closing the plant nutrient loop. However, when it comes to their issued rules, many of them place urine and black water in the same category as sewage sludge and they do not buy farm produce which has been fertilized with it. The reason for this lack of dedicated rules for urine and faeces is that the volume of recycled urine and black water is so far too small to motivate the companies to go through the cumbersome procedure of adopting special rules for these recycled fertilizers. This meant that urine and black water mainly was used for fodder and energy crops by the farmers interviewed.

Another restriction was the need to sanitize the collected products before they were used as fertilizers. This meant that collection had to be well planned in relation to the time of fertilization. There had to be sufficient time in the storage for the urine and black water, so that they had time to sanitize sufficiently in relation to the crops they were used for. In the Stockholm and Håga cases, this was solved by their access to several storage tanks, making it possible to spread only the urine from those tanks, which had been stored sufficiently long. In the Lund and Norrköping cases, it was solved by the transport from the collection tanks to the storage tanks being scheduled to the autumn to ensure sufficient storage before the spreading in the spring and summer. For the Kvicksund and sewage sludge cases, the treated and delivered products were sufficiently sanitized and could be spread directly after the delivery/treatment. In the case of Tanum, it was handled differently between different farmers. Frequently the urine was spread directly from the collection tanks before ploughing and establishing a new crop. In this way, the transport and handling was minimized.

11.5 Farmers as Entrepreneurs for Handling and Reuse

An important driver for several of the farmers involved in the recycling schemes was that this was a new concept, where they saw new business opportunities. Farmers are largely entrepreneurs and businessmen. During the year, there are several periods with less work and many farmers are looking for profitable tasks to fill these periods.

Thus, several of the studied farmers were interested in carrying out, for a fee, the collection and storage of the sanitation products to be recycled, in addition to spreading and utilizing them as fertilizer. In two of the recycling schemes studied (Håga & Tanum), the farmer was responsible for the collection and transport and for these farmers the payment for these services was a very important incentive.

Involving the farmer as an entrepreneur is also good, as the business opportunity often overrides the possible scepticism towards the waste water fractions, which some farmers have: "We do not want to be a dumping station for waste from society". By involving the farmer as a business partner at an early stage, sound business decisions will often outweigh this scepticism.

For the sustainability of the recycling system, using the farmer as an entrepreneur for handling is also interesting as this will decrease the number of actors involved in the organization and increase the interaction between the farmer and the other actors in the chain. It provides a valuable direct contact between the source separating household and the farmer recycling the product, and thus a possibility for direct feedback on how the source separated product is used. Furthermore, simulations have shown that slurry spreaders are more energy efficient for the collection and transport, over reasonable distances, than suction trucks.

When a municipality contracts entrepreneurs for the collection of household waste, such as urine and black water, this is preceded by a standardized tender process. For the farmers to be competitive in this process, it has to be changed in such a way that small enterprises can qualify to deliver the service. The farmers also need education on how to produce a good tender and on the legal aspects of collecting and handling household waste.

A technical aspect that often turns up when planning storage and treatment is that existing slurry tanks in agriculture can be used to store and treat urine and black water. These tanks are however often quite large in comparison with the amount of urine and faeces that can be recycled at early stages. Furthermore, they almost always lack lids and there is thus often a need to equip the tank with a lid in order to minimize ammonia losses.

11.6 Recommendations

11.6.1 Include the Farmers in the Initial Planning

For a successful recycling system, the farmer recycling the product is as essential as the household where the product is collected from. For the success of the system, it is essential to involve the farmer, right from the initial planning of a recycling system. Preferably the municipality cooperates with interested farmers before the recycling systems are introduced in the municipality, as knowing the possibilities and preferences of the farmers can have a decisive influence on the organization of the system, and thus greatly improve its efficiency and sustainability. If for example the farmer has an unused slurry basin, it might be possible to use this for storage and treatment, as was the case in Norrköping, Tanum and Lund. If the farmer is interested in carrying out the collection, then this can simplify the system and decrease the costs, as was the case in Tanum. Furthermore, by including the farmer early in the planning, there will be more time for the different stakeholders to understand each other and to gather around similar values.

Most importantly, by including the farmer in the planning from the beginning, chances are maximized that the products will be recycled, which in turn has proven to be very important for the sustainability of the rest of the system. In a number of Swedish systems the urine diversion toilets have been replaced with conventional flush toilets. One common trait for all of these, save one or two, is that the diverted urine has not been recycled, but has overflowed to the conventional sewage system. It is not surprising that the households in such systems after some years become tired of the extra maintenance of urine diversion toilets and resume using conventional toilets.

11.6.2 Organize an Arena for Feed-Back and Interaction

Initially, when the system is organized and started up, there is usually a lot of interaction between the different stakeholders. However, once the system is running, this interaction easily dies. Recycling systems for sanitation products are new and their improvement potential is large. To explore and utilize this potential, as well as to solve possible problems, it is important that the stakeholders have an arena where they all can meet and interact (Storbjörk and Söderberg 2003). In Tanum, the stakeholders met once a year for feed back and exchange of experiences. These meetings were much appreciated and considered important for the further improvement of the system, and for the stakeholders to understand each other and to share a common vision for the system.

11.7 Conclusions

In a recycling sanitation system, the farmers recycling sanitation products are just as important as the households producing them. It is important that farmers are involved and take part in the planning right from the beginning.

There are many stakeholders in a recycling sanitation system and it is important that the drivers and restrictions of each stakeholder are understood. This is especially important concerning the farmers, as they are key stakeholder in the proposed recycling systems. The knowledge on their drivers and restrictions is still very low. Farmers are businessmen and the recycling system can often become more sustainable if the farmers are also conceived of as entrepreneurs for the collection and handling of the products, as this enhances the utilization of their business potential.

It is also important that there is an arena where the different stakeholders of the system meet and communicate. This forum is also needed for maintaining a shared vision of the system between the stakeholders, which is crucial for reaching the goals of the system. The coordinator needs to know farming and the conditions for the farmers sufficiently well to communicate with them effectively.

References

Aubain, P., Gazzo, A., Le Moux, J., & Mugnier, E. (2002). *Disposal and recycling routes for sewage sludge – Synthesis report*. European Commission DG Environment-B/2.

Barles, S. (2007). Feeding the city: Food consumption and flow of nitrogen, Paris, 1801–1914. *Science of the Total Environment, 375*, 48–58.

Bengtsson, M., Lundin, M., & Molander, S. (1997). *Life cycle assessment of wastewater systems. Case studies of conventional treatment, urine sorting and liquid composting in three Swedish municipalities*. Report 1997:9, Technical Environmental Planning, Chalmers University of Technology, Sweden. Retrieved from www.chalmers.se

Jönsson, H. (2002). Urine separating sewage systems – Environmental effects and resource usage. *Water Science and Technology, 46*(6–7), 333–340.

Jönsson, H., Ashbolt, N., Baky, A., Drangert, J.-O., Krantz, H., Kärrman, E., et al. (2005). *Slutrapport från modellstaden Urbana enklaven* (Final report from the model city the Urban Enclave; in Swedish). Report 2005:8, Urban Water, Chalmers, Sweden. Retrieved from www.urbanwater.org

Storbjörk, S., & Söderberg, H. (2003). *Plötsligt händer det. Institutionella förutsättningar för uthålliga va-system* (Suddenly it happens. Institutional conditions for sustainable water and sanitation systems. in Swedish.) Urban Water Rapport 2003:1, Chalmers, Sweden. Retrieved from www.urbanwater.org

Tidåker, P. (2007). *Integrating farming and wastewater management*. Doctoral dissertion, Department of Biometry and Engineering, Swedish University of Agricultural Sciences. *Acta Universitatis Agriculturae Sueciae, 2007*, 85.

Tidåker, P., Kärrman, E., Baky, A., & Jönsson, H. (2006). Wastewater management integrated with farming – An environmental systems analysis of a Swedish country town. *Resources, Conservation and Recycling, 47*(4), 295–314.

Tidåker, P., Mattsson, B., & Jönsson, H. (2007a). Environmental impact of wheat production using human urine and mineral fertilisers – A scenario study. *Journal of Cleaner Production, 15*, 52–62.

Tidåker, P., Sjöberg, C., & Jönsson, H. (2007b). Local recycling of plant nutrients from small-scale wastewater systems to farmland – A Swedish scenario study. *Resources, Conservation and Recycling, 49*, 388–405.

Tidåker, P., Sjöberg, C., & Jönsson, H. (2007c). Local recycling of plant nutrients from small-scale wastewater systems to farmland – A Swedish scenario study. *Resources, Conservation and Recycling, 49*, 388–405.

Wetterberg, O., & Axelsson, G. (1995). *Smutsguld och dödligt hot* (Dirty gold and deadly threat. In Swedish). Sweden: Chalmers University of Technology.

Chapter 12
Governing Peri-Urban Waste Water Used by Farmers: Implications for Design and Management

Reginald Grendelman and Frans Huibers

Abstract Worldwide, population is increasingly centralized in metropolitan areas. This has an impact on water systems and complex metropolitan watersheds emerge. Flows of varying water quality are generated and distributed among different users who develop new opportunities and coping mechanisms for dealing with marginal quality waters. In developing countries waste water management often fails to cope with the increasing number and volumes of flows. Financial and institutional limitations force waste water managers to discharge substantial amounts of untreated or partially treated waste water into surface waters. Consequently, use of polluted water is increasingly common in the downstream peri-urban agricultural areas. This, albeit productive, may lead to negative impacts on human health and environment, if management of this water is not rightly done. Mitigation of the problems requires rethinking of conventional 'top-down' waste water system design and management, in combination with expected down-stream use. In this chapter the applicability of water governance principles in design and operation of waste water systems with an effluent use component is investigated. Acknowledgement of the treatment potential of subsequent uses and the significance of use-based practices as opposed to zero-pollution design will certainly change design and treatment procedures. Inclusion of agriculture and nature as a treatment step and participation of users in decision-making are expected to optimize use of finances, infrastructure and personnel.

12.1 Introduction

In urbanizing areas surrounding the world's larger cities growing disparities become visible between waste water management and daily practice of waste water use. This chapter investigates how and by what factors management and practices are shaped. It assesses the applicability of introducing governance principles in

R. Grendelman and F. Huibers (✉)
Irrigation and Water Engineering Group, Wageningen University, P.O. Box 47,
6700 AA Wageningen, The Netherlands
e-mail: reginaldgrendelman@hotmal.com; frans.huibers@wur.nl

B. van Vliet et al. (eds.), *Social Perspectives on the Sanitation Challenge*,
DOI 10.1007/978-90-481-3721-3_12,

waste water management in which downstream recipients are involved in decisions on design and management.

First, in the remainder of the introduction, the complexity of peri-urban waste water management is mapped. Next the current state of waste water use with its challenges and limitations is sketched out. Alternative approaches to waste water management like semi-centralized systems and the waste water chain are then discussed and the concept of governance as applied to waste water management is introduced. In the following sections the boundary conditions of applying waste water governance in peri-urban areas are described and from this the implications for the design of waste water facilities and institutions based on empirical findings are put forth. In the closing section some conclusions are presented for discussion.

Conventional management of waste water in urban areas is most generally based on collection and transport of waste in water-fed sewers to centralized treatment plants. Advantages of this system are, among others, the low risk of contamination thanks to a closed network of pipes, assumed economy of scale for the treatment system and the quality controlled outlet of effluent to feed back into the water system. However, rapidly expanding cities in developing countries require ever-longer distances of sewerage lines. Both investment and costs for operation and maintenance are high and lengthy construction time prevents timely reaction to rapid demographic changes. Moreover, such conventional system is based on the assumption that available treatment capacity is sufficient to treat all waste water inflow up to an acceptable and agreed upon quality level.

Designed for the gradually growing major cities in the nineteenth and early twentieth century, the centralized approach has posed little problems for waste water managers. Economic growth kept pace with the increase in population and the sewer networks were steadily extended. In contemporary cities in developing countries the situation is very different, growing at high pace and soon housing 80% of the five billion urban people. Such explosive growth of the number of inhabitants exceeds the economic growth. This results in a disability of urban waste water managers to cope with the growing flows of waste water due to insufficient institutional and financial capacity. Besides this, as Marcotullio (2007) describes for cities in Southeast Asia, a what he calls *time-space telescoping* of development is taking place: in contrast with the sequential and staged development of cities in the now developed world current development is taking place simultaneously at different scales within urban areas, leading to sets of environmental problems occurring at once (see also Oosterveer et al., Chapter 2 of this volume). Numerous problems on pollution, poverty, food security, water supply and sanitation, especially in the often unplanned and informal outer parts of the urban centres can be contributed to these phenomena (Varis et al. 2006; Biswas 2006; Marcotullio 2007).

Peri-urban areas can be seen as places that develop from rural to areas with distinctive influence by the neighboring city. The city's influence shows in different forms: negative where it concerns the increasing pollution of the surface water, positive where the city develops as an interesting market offering new opportunities for the farmers. Other changes occur as well, like immigration and emigration of labor, changing traditional village social structures, leading to fuzzy institutional structures and unclear responsibilities. According to Ducrot et al. (2007) migration and high mobility result in absence of formal institutions and weakened leadership in peri-urban zones. To these problems, Drechsel et al. (1999) add that insecurities in tenure lead to short-term planning and unsustainable land-use and prevent investments in, for instance, waste water use facilities.

Being downstream, the peri-urban areas are a major receiver of the urban waste water. Volumes of (treated) waste water discharged from metropolitan cities can reach levels of 40–60 l/c/day. For example, a city with a five million population would then discharge 250,000 m^3/day, equivalent to over 90×10^6 m^3/year. Such volume could potentially irrigate 9,000 ha of land at 1,000 mm/ha/year, depending on agronomic conditions (soil, climate, crop choice) and irrigation techniques and water management. Although farmers, if they would have the choice, would generally prefer fresh water to irrigate, such waste water flows are interesting for them, giving their permanency and reliability, in which some formal irrigation schemes fall short. Farmers could also beneficially use the nutrients (N, P and K) in the (treated) waste water, although a number of conditions should be met, including proper communication and information on the water quality. Too high levels of nutrients would have negative effects on yields and be a source of further downstream pollution. Toxic elements should be absent, as they could enter the food chain, while presence of pathogens would require careful handling of water in the field to protect field laborers and careful handling of the crops to protect consumers, even more in the case of crops that are consumed raw.

12.2 Agricultural Use of Waste Water

The use of urban waste water in agriculture has been practiced worldwide for ages. Due to problems with public health related to exposure of crops to pathogens, the need for residential area near the cities and the rise of new treatment technologies for waste water the sewage farms, as described by Rafter (1897) were stripped down (Van Loohuizen 2006) and replaced by centralized treatment plants.

From the 1950s onward the use of domestic waste water in agriculture has gained renewed interest, specifically in areas of high water stress, like the Middle East. In more recent years the use of waste water is increasingly described and recognized as a reality, being the consequence of uncontrolled discharge of (treated) waste water in otherwise clean water sources in combination with the sometimes desperate need for water by the farmers. Officially, the use of untreated or partly treated waste water in irrigated agriculture is forbidden in many countries. However, the contribution to local food markets of crops produced in peri-urban agriculture using waste water is substantial. Smit and Nasr (1992) estimated that already that time around 10% of the world population consumed food, which was irrigated with (partially treated) waste water. In the last decades this figure is expected to have grown. However, data on the extent of waste water use for irrigation is scarce. Scott et al. (2004) note an estimate of 20 million hectare of waste water irrigated land, including the use of partially treated waste water. They argue that waste water use is likely to grow as sanitation capacities in the coming years will probably fail to keep pace with the increase in water supply. The expansion of waste water irrigation in peri-urban areas can be contributed to a number of factors. Urban growth and wealth result in an increase in water consumption. Increased demand for vegetables by population growth results in year round cropping, even in dry periods when the natural water is absent. Contrary to common believe, waste water use in peri-urban areas is not restricted to (semi-)arid regions. Also the seasonal shortage of fresh water and reliable access to waste water results in waste water irrigation in metropolitan areas in more humid areas.

Waste water use in agriculture introduces risks and opportunities for agricultural water management. Hussain et al. (2002) identified different impacts when waste water is used for irrigation. A major concern is the negative impact waste water can have on public health. The presence of pathogenic microorganisms, and especially human parasites like protozoa and helminth eggs, pose threats for the emergence of waterborne diseases where untreated waste water is used. For instance, the prevalence of hookworm and *Ascariasias* infections is higher among children where waste water irrigation is practiced (Hussain et al. 2002). The presence of heavy metals can present a risk for irrigators through skin contact. Crops generally do not take up such toxics, although few do, this way posing a risk to consumers. Accumulation of toxics in the food chain can also start through cattle grazing in the irrigated fields. Impacts on crops can be both beneficial and harmful to crop growth. Nutrient availability, especially nitrogen (N), phosphorus (P) and potassium (K), results in increased yields and lower fertilizer requirements. The relative constant supply of waste water extends growth periods and may increase the number of cropping seasons in 1 year. However, overloading with nutrients will reduce yields in grain crops or result in delay of ripening and maturity. Accumulation of waste water constituents in soils is often problematic. The addition of nutrients (N, P), suspended solids, heavy metals and especially salts pose threats to the long-term agricultural productivity and may negatively affect the soil structure. Increased soil salinity and nutrient content can lead to contamination of groundwater by leaching processes. The drained waste water can also increase the load of pathogenic bacteria and viruses in groundwater, which is problematic since in many areas groundwater is used for domestic purposes (NRC 1996 in Hussain et al. 2002). For these factors guidelines exist stressing the maximum acceptable loads in waste water, but Toze (2005) mentions that, although prevalent and potentially dangerous, trace elements, disinfection-by-products and pharmaceutically active compounds are rarely incorporated in guidelines for waste water use.

Acknowledgement of the fact that waste water is often used unregulated requires rethinking of effluent and surface water quality. This downstream use in agriculture or in nature needs specific quality levels of the effluent. When considering agricultural use the crop choice, cropping pattern, irrigation method and technology, as well as farming practices should be considered, in combination with actual water quality.

12.3 Alternative Approaches to Waste Water Management

Traditionally waste water management is the domain of environmental engineers who develop essentially technical solutions for problems posed by waste water flows. Designs are mostly steered by technical parameters like treatment efficiency and drainage capacities. Social implications are often confined to public health and environment. Socio-economic benefits which are directly related to waste water use like food and income security are less considered in waste water management. Ignoring this reality results in a lack of involvement of peri-urban farmers in considering options for facility design. These downstream water users are currently not incorporated in the decision-making, leading to uncertainty on water quality and quantity which in turn leads to livelihood insecurity.

Along with increasing attention for the actual use of waste water, especially in developing countries, and better understanding of its possibilities and constraints, a growing number of authors support new technologies and management approaches to handle waste water flows. In this context decentralization in the collection and treatment of domestic waste water is often suggested. In the extreme, household-based treatment systems can prevent the high investment costs in sewerage infrastructure although safeguarding hygiene and water quality is difficult, especially in the densely populated metropolitan areas.

Weber et al. (2007) propose the use of semi-centralized supply and treatment systems (SESATS) where the benefits of short strings of infrastructure are combined with the possibility to have good quality control. SESATS are furthermore helpful to separate domestic and industrial flows. This allows treatment to be better aimed at the constituents and enables easier use of the effluent (Tandraatmadja et al. 2005).

As useful conceptual framework Huibers and Van Lier (2005) proposed to consider the water chain where "the water is followed along the path from originally fresh water made available for high value domestic or industrial use, leaving these activities as waste water, being upgraded by treatment facilities and subsequently brought to agricultural fields". In a further contribution (Van Lier and Huibers 2007) the same authors suggest that an optimal design of a waste water treatment system could be realized if design parameters are based on downstream use of the effluent. If this were irrigated agriculture, one would, for example, not aim to remove (all) nutrients from the waste water as they could be beneficially used in agriculture.

Evers et al. (2008) describe the application of supply chain theory to describe the interactions along the line of stakeholders manipulating the water source. It aims to provide a socio-technical approach combining the technical water chain approach "based on material flows and social scientific approaches based on actor networks and institutional organization". Martijn and Huibers (2001) identified a number of decision-making items for the technical design of treatment systems with effluent use. These would include type of post-treatment, effluent supply, irrigation method and technology, crops and farming practices. Depending on these items, irrigated agriculture would be able to receive and use different qualities of water. Irrigation itself would indeed be a treatment step before draining the water for further destinations. Such design approach allows developing strategies for water treatment and handling making optimal use of possibilities, scopes and room for maneuver of all involved stakeholders combined with their wishes and needs.

To facilitate such a design approach it is important that information regarding sustainable and safe use of waste water is available to all involved and that public perceptions on the issue reflect reality rather than presumptions and local convictions. Clarity in information can be vital in negotiation between stakeholders, legal disputes regarding water rights and the erection of solid, demand-based waste water management to sustain (peri-) urban livelihoods and environment.

Having the potential to absorb at least part of the pollutants, agriculture (and also aquaculture which is not further discussed here) should be included in the designing of waste water treatment systems. This allows incorporating post-treatment like nutrient removal and enables to make waste water systems more specifically targeted at separation or removal of toxic and hazardous substances like pathogens or trace-metals. Insights in seasonal variation in nutrient demand of crops might make variable and (from a nutrient perspective) more efficient treatment at the sewage plant possible. This, however, requires a more sophisticated treatment approach. In line with such bottom-up approach, the social structure of peri-urban farmers should be strengthened by forming stakeholder groups who negotiate the desired water quality with waste water treatment managers at city-level. This also forces local authorities to acknowledge that waste water at different levels of treatment is used in peri-urban irrigated agriculture, providing ground for realistic policies and equitable division of water rights among different uses and users within and outside the city.

With an eye on the complexity of social, institutional, hydrological and environmental processes in waste water management with a use component governance theory is introduced here as a potentially fitting concept. Governance theories aim to describe networks of interdependent actors and their rule-making and steering mechanisms, processes by which rules about the pursuance of public goods are designed and enforced. Governance is about affecting 'the frameworks within which citizens and officials act' (Kjaer 2004). Water governance is a well-researched and much debated element of irrigation water management. The analyses and discussions are in that sector focused on demand and supply issues and questions of how to include local stakeholders and downstream users at different levels of decision-making and management of the available water resources. Although the concept of governance in waste water management is not fully developed, basic principles of governance do certainly apply to this institutional and socio-economical complexity, with elements of rules and technology choices, planning and management, different interests between actors and unclear boundaries between public, private and associative spheres.

12.4 Waste Water Governance in Peri-Urban Areas

In the peri-urban fringes of cities the complexity of waste water management is most visible. Here waste water use is an issue and public acceptance and perceptions are often greatly affected by the way and extent in which waste water is used.

In this context a conventional top-down approach of waste water facility design seems hardly viable.

This calls for a different approach in which the productive use of waste water and downstream environmental effects are included as part of the solution for dealing with waste water. Given the characteristics of the peri-urban areas, the complexity of waste water and the variety of waste water use, the application of water governance principles on waste water management seems a feasible option to minimize negative social and environmental impacts. Governance of resources starts with the determination of agreed upon objectives between stakeholders. For waste water a number of objectives could easily be defined, like environmental and public health protection and maintaining food security or poverty alleviation (WHO 2006). In order to set and reach the objectives a participatory process seems most suitable. This requires, among others, institutionalization of waste water management, increasing community acceptance and social strengthening.

Institutionalization of waste water use is not common. In water scarce countries competition between different water sectors may stress cooperation which would be needed when policies on waste water are to be implemented effectively. Changes in the institutional approach of payment for water treatment, for example, require rethinking of which stakeholders bear which costs and who sees profit of the treatment. If any waste water use initiatives are in place they are often conducted by fragmented and unconnected institutions, resulting in locally independent projects lacking overview needed for a sustainable design of waste water facilities (Neubert 2002).

Currently, waste water treatment is often institutionalized in classic setups. Centralized governments, in most cases urban officials, are responsible for the operation of the waste water collection and treatment. Based on data on flows and quality of the flows reaching the treatment plant and the given standards for the plant's effluent the optimal treatment methods are determined. Panebianco and Pahl-Wostl (2006) criticize the absence of the human dimension in conventional systems. The human dimension is seen as an exogenous boundary condition and not taken into account in the planning and design processes. They see existing treatment systems as socio-technically constructed units, which have evolved over time. Acknowledgement of the human dimension should lead to changes in the institutional set-up and would introduce a need for participatory design and operation for optimal results. Part of the human dimension is the degree of acceptance for the use of waste water at community level, which depends on many factors. In many developing countries peri-urban farmers have no choice, as waste water (often in the form of polluted surface water) is their only water source. In developed countries waste water use depends strongly on acceptance. In general, the level of acceptance increases when contact with, or proximity to, waste water decreases. Irrigation of landscapes is thus better accepted than irrigation of food crops (Hamilton et al. 2007). Public unease is furthermore affected by the source of waste water. For instance, untreated storm water runoff, although in some cases as potentially dangerous as partially treated effluent, is often perceived as less threatening to public health than the effluent (Toze 2005). Kahn and Gerrard (2006) identified that, besides proximity and source, trust of the water users in the institutions dealing with waste water is important. This trust is based on the credibility of the institution and its cooperation with the community. The credibility of an organization is perceived to be higher when it is committed to the welfare of the community, is impartial, knowledgeable and experienced. Cooperation between the users and the institutions is best when the community is actively involved from an early stage with the possibility of delivering ideas, opinions and having control in the process. Within this communication open access to unbiased information is important. Increasing knowledge and understanding of the principles of waste water use generally lead to better acceptance of waste water use projects by (relative) laymen such as farmers. Next to increased acceptance, training and dissemination of information is necessary for better performances in operation and maintenance of the system (Parkinson and Tayler 2003).

A tool for introducing the human dimension in waste water management is the setting up of waste water management teams in which all stakeholders are represented. Although often coined as the key element of stakeholder participation, one should realize that multi-stakeholder platforms may not always result in a situation of perfect communication and social learning. Stakeholders are tempted to pursuit personal gains and the process is in many cases politicized. To prevent the waste water management teams of becoming instruments of the already powerful to take over management of waste water a level playing field is desired, although difficult to implement. Capacitating and empowering of the weaker stakeholders should at least be secured at the very start of participatory processes. Also forming alliances between the weaker stakeholders should be allowed when interests collide (Faysse 2006).

Within waste water management teams developments within the peri-urban area can be discussed and their impact on waste water practices determined, because, especially in "a situation of unclear data, stakeholders find joint fact-finding, exchange and relations useful and, in several cases, enjoyable" (Warner 2006). Therefore, after inception and empowerment of weaker stakeholders, waste water management teams should facilitate a focus group of which the exchange of knowledge and experiences on waste water use and its management are central elements. In order to enhance mutual understanding of problems and degrees of freedom to maneuver in, role-playing workshops provide useful tools for minimizing negative effects by increasing understanding of resource dynamics and creating a common sense of problem ownership. From this point a vision and strategy can be drafted for future management, rooted in the local situation.

It is important that these teams are not seen as the answer to a lack of human dimension, but merely as a means to enhance participation and to open up the debate for all involved. Warner (2006) states that "… just sitting together does not solve problems. People have to bring, or develop, skills for making a multi-stakeholder process work, and need to keep catering to stakeholders' immediate needs and interests to 'reproduce' the platform, accepting that not everybody will participate." This would advocate for an approach in which involvement is gradually scaled up from knowledge exchange through vision building to co-management. From the beginning special attention should be paid to power relations within and composition of the team and the representation and participation capacity of the involved actors. The decision-making powers and mechanisms should be clearly defined and the costs of setting up such a team and the longer project running time due to negotiations should be taken into account (Faysse 2006).

12.5 Implications for the Design of Treatment Facilities

Treatment based on downstream demands requires other technical criteria than conventional treatment, where population and loadings, industrial discharges and water quality standards are basic inputs for process design (ASCE 1998). This is depicted in Fig. 12.1, where population and loadings and industrial discharges are grouped as waste water characteristics. When downstream use is considered as a determining factor for the effluent quality of a treatment plant it is no longer realistic to aim for a constant quality of effluent. For instance, the fluctuation of nutrient demand due to different requirements of the crops in different stages of plant growth results in a need for varying offering of nutrients. Crop water requirements vary depending on the different growth stages and thus result in demands for more or less flows of water. Figure 12.2 gives an overview of a design scheme based on effluent use.

Besides agricultural consideration on the demand side, it is necessary to allow differentiation at the supply side. Many crops are intolerant for high salinity levels and certain types of heavy metals and toxins. The latter are generally stemming from industrial water use. By separating domestic and industrial waste water flows it is possible to increase the use potential of the waste water flow. Inclusion of downstream farmers in the treatment chain provides a means to introduce additional payment for services with these farmers. Where currently waste water is

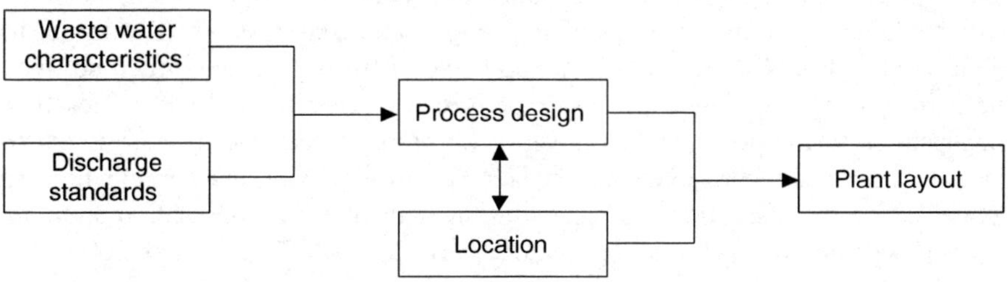

Fig. 12.1 Conventional design scheme

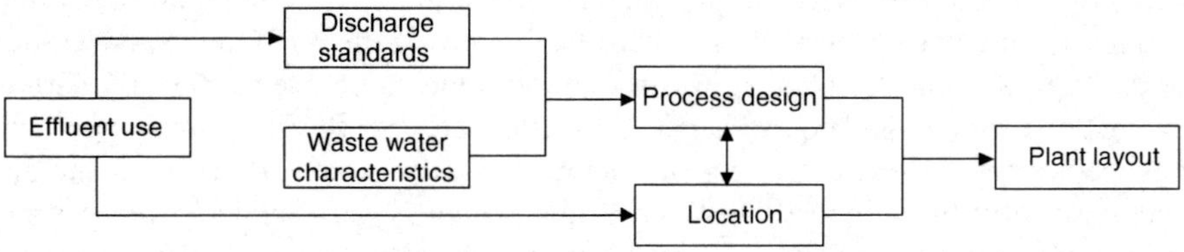

Fig. 12.2 Design scheme based on effluent use

regarded as a residual product the presence of nutrients, and the year-round availability make it actually a precious resource, especially in arid regions. With a flow tailored to their needs, farmers are able to achieve higher yields due to well-tuned nutrient supply and more crops per year. If this leads to sustainable additional profits it might be possible to ask compensation from the farmers, thus creating an additional financial basis for waste water treatment. In coordination with the waste water management teams the boundary conditions of planned treatment facilities should be revisited. For instance, the envisaged crops in the area determine the sensibility to salt, trace-elements or heavy metals, as well as the crop water requirement and nutrient demands.

Another seemingly 'standard' boundary condition is the location of the treatment facility. Traditionally the treatment plant is located as low and downstream as possible to facilitate gravity flow of the waste water. When, however, the effluent is used in agriculture the location is preferred as upstream as possible to allow the waste water to flow by gravity to the irrigated fields. Hence a trade-off is necessary, depending on more location specific features than ground elevation alone.

12.6 Implications for Waste Water Institutions

Introducing water governance principles in waste water treatment systems has a number of institutional implications. To include the downstream farmers in design and decision-making it is crucial they are grouped together in local organisations with sufficient knowledge on the locality, agricultural practices and environmental engineering. The government institutions in place have to provide technical assistance and capacity strengthening, coordinate activities and negotiations and develop monitoring and regulation systems rather than imposing centralized systems (Parkinson and Tayler 2003). The Negowat project in São Paulo in Brazil revealed difficulties when the participatory negotiation processes of land use planning and natural resources management were initiated. Most government professionals had little or no experience in participatory processes and stakeholder involvement and participation was conceived by the institutions as presenting their plans to the population and afterwards proceeding in a paternalistic and one-way approach (Ducrot et al. 2007).

Yet, although not formally initiated, several cases can be referred to in which the actual downstream use of waste water has influenced upstream decision-making, breaking with the conventional design procedure and decision-making. In Mexico City downstream farmers protested against new policies to remove nutrients they had been using in crop growing for considerable time, which led to reconsideration of the suggested master plan for waste water treatment (Jiménez 2005). By harmonising downstream use requirements and waste water treatment options opportunities were created to treat a larger portion of the waste water on the one hand and keeping useful nutrients in the waste water, which is beneficial for waste water irrigated agriculture. In Dakar, Senegal, part of the waste water from the city is used in peri-urban agriculture, mostly without any form of treatment. Farmers organized themselves to resist the enforcement of rules as no other water source would be available for them. They also actively dispute urban land developments that would expropriate their area (Redwood 2009).

Another important institutional implication of a use-based design is the need to abandon the strict discharge limits and effluent quality criteria, which are currently often decreed at national or international level. When incorporating agricultural use of the waste water the output of the treatment plant will inherently need to become more variable in terms of quantity and quality. Seasonal crop water and nutrient requirements prevent the discharge of constant flows of uniform quality. Furthermore crop choice may determine varying requirements for quantity and quality for different discharge areas. Currently crop restrictions, such as in Tunisia, limit the production of market crops, hampering the economic development of effluent use (Bahri and Brissaud 1996).

A well documented case is Faisalabad, Pakistan, where some farmers, with access to treated and untreated waste water, opted for the untreated waste water as it was considered less saline and better for their crops (Ensink et al. 2004). The case reported by Ratner and Gutiérrez (2004) on the Guatamalan highlands shows how a waste water treatment plant was constructed without considering local institutions. In this case operation and maintenance of the system proved problematic due to lack of involvement but the importance of the plant was recognized by all and this opened avenues for public dialogue and community action, leading to joint action. Lastly, Evers (2006) and Evers et al. (2008) describe that *vertical integration*, as defined in the supply chain theory, is an essential element of waste water management based on the final destination or ultimate use. This integration can be achieved by better communication among the entire waste water chain creating a common goal. This sets the environmental engineers out to develop communication strategies in which impacts of decisions are clearly communicated, with their respective impacts on the system and, even more importantly on the stakeholders themselves.

The examples given in this section have in common that they were based on farmer's protest against government planning, in which their own stake was not or insufficiently considered. They also show that adaptations in the waste water management proved feasible and proved beneficial. It stresses the need for inclusion of all relevant stakeholders like water users and specialists from multiple technical and social disciplines who acknowledge "the many facets of engineering solutions to problems [in developing communities], beyond the [technical] skills obtained in their basic education". This helps to set up sustainable projects which are "symbiotic with the environment, society and culture and help build capacity for people to solve their own problems" (Robbins 2007).

12.7 Discussion

Introduction of water governance principles in design and operation of waste water systems has a number of consequences for environmental engineers. Regarding their technical background and extensive experience with all sorts of alternative treatment methods and their requirements the technological consequences are not expected to be insurmountable. The inclusion of farmers in design and operation and truly participatory approaches might prove to be difficult and it is important that knowledge on agricultural and natural demands and constraints is introduced in the waste water management teams. This can be done by employment or by cooperation with other local governments and experts. The latter is more sustainable as it promotes interaction between different sectors and levels of management. To this end the establishment of negotiation platforms, or waste water management teams, in which all stakeholders have a say in design and operation and maintenance practices can prove useful. However, negotiation processes can be lengthy and the waste water management teams have to be developed gradually. Governing of waste water treatment systems may thus take more time to come to an agreed-upon design, stretching the planning process. Furthermore, negotiation strategies with hidden agenda's may obscure the processes. For environmental engineers this means that their profession is changing from finding solutions to technical problems to finding solutions to socio-technical problems. Their role changes from prescriptive engineers to reflexive engineers (Robbins 2007) and mediators in negotiation processes.

References

ASCE. (1998). *Design of municipal wastewater treatment plants* (4th ed., Vol. 1). Water Environment Federation manual of practice no.8 and American Society of Civil Engineers manual and report on engineering practice no. 76. Alexandria: WEF; Reston, VA: ASCE.

Bahri, A. & Brissaud, F. (1996). Wastewater reuse in Tunisia: Assessing a national policy. *Water Science Technology, 33*(10–11), 87–94.

Biswas, A. K. (2006). Water management for major urban centres. *International Journal of Water Resources Development, 22*(2), 183–197.

Drechsel, P., Quansah, C., & Penning De Vries, F. (1999). Urban and peri-urban agriculture in West Africa. In O. B. Smith (Ed.), *Urban agriculture in West-Africa – contribution to food security and urban sanitation*. Ottawa, Canada: International Development Research Centre.

Ducrot, R., Chagas de Carvalho, Y. M., Jacobi, P. R., Clavel, L., Barban, V., Madazio, V., et al. (2007). Building capacities to tackle the infrastructural and environmental crisis in São Paulo: Role-playing games for participatory modelling. In J. Butterworth, R. Ducrot, N. Faysse, & S. Janakarajan (Eds.), *Peri-urban water conflicts. Supporting dialogue and negotiation*. Delft, The Netherlands: IRC International Water and Sanitation Centre.

Ensink, J. H. J., Simmons, R. W., & Van der Hoek, W. (2004). Wastewater use in Pakistan: The cases of Haroonabad and Faisalabad. In C. A. Scott, N. I. Faruqui, & L. Raschid-Sally (Eds.), *Wastewater use in irrigated agriculture – coordinating the livelihood and environmental realities*. UK: CAB International, International Water Management Institute and International Development Research Centre.

Evers, J. G. (2006). *Everybody's business is nobody's business. A study on the use of wastewater in (peri-) urban irrigated agriculture in Hanoi, Vietnam*. Unpublished M.Sc. thesis, Irrigation and Water Engineering Group, Environmental Policy Group, Wageningen University, The Netherlands.

Evers, J. G., Huibers, F. P., & Van Vliet, B. J. M. (2008). Institutional aspects of integrating irrigation into urban wastewater management: The case of Hanoi. *Irrigation and Drainage*, Online first, doi: 10.1002/ird.466.

Faysse, N. (2006). Troubles on the way: An analysis of the challenges faced by multi-stakeholder platforms. *Natural Resources Forum, 30*, 219–229.

Hamilton, A. J., Stagnitti, F., Xiong, X., Kredil, S. L., Benke, K. K., & Maher, P. (2007). Wastewater irrigation: The state off play. *Vadose Zone Journal, 6*(4), 823–840.

Huibers, F. P. & Van Lier, J. B. (2005). Use of wastewater in agriculture: The water chain approach. *Irrigation and Drainage, 54*, S3–S9.

Hussain, I., Raschid, L., Hanjra, M. A., Marikar, F., & Van der Hoek, W. (2002). *Wastewater use in agriculture: Review of impacts and methodological issues in valuing impacts*. Working paper 37. Colombo, Sri Lanka: International Water Management Institute.

Jiménez, B. (2005). Treatment technology and standards for agricultural wastewater reuse: A case study in Mexico. *Irrigation and Drainage, 54*, S23–S33.

Kahn, S. J. & Gerrard, L. E. (2006). Stakeholder communications for successful wastewater reuse operations. *Desalinisation, 187*, 191–202.

Kjaer, A. M. (Ed.) (2004). Introduction: The meaning of governance. *Governance*. Book in the Key Concepts series. New York: Wiley.

Marcotullio, P. J. (2007). Urban water-related environmental transition in Southeast Asia. *Sustainability Science, 2*, 27–54.

Martijn, E. J. & Huibers, F. P. (2001). *Use of treated wastewater in irrigated agriculture. A design framework*. Wageningen: Coretech.

Neubert, S. (2002). *Wastewater reuse in agriculture – A challenge for administrative coordination and implementation*. In S. Neubert, W. Scheumann, & A. van Edig (Eds.), *Reforming institutions for sustainable water management*. Reports and Working Papers 6. Bonn: German Development Institute.

NRC. (1996). Use *of reclaimed water and sludge in food crop production*. National Research Council. Washington, DC: National Academy Press.

Panebianco, S. & Pahl-Wostl, C. (2006). Modelling socio-technical transformations in wastewater treatment – A methodological proposal. *Technovation, 26*, 1090–1100.

Parkinson, J. & Tayler, K. (2003). Decentralized wastewater management in peri-urban areas in low-income countries. *Environment and Urbanisation, 15*, 75–91.

Rafter, G. W. (1897). *Sewage irrigation. United States Geological Survey* (100 pp.). Water Supply and Irrigation Papers No. 3.

Ratner, B. D. & Gutiérrez, A. R. (2004). Reasserting community: The social challenge of wastewater management in Panajachel, Guatamala. *Human Organization, 63*(1), 47–56.

Redwood, M. (2009). *Agriculture in urban planning. Generating livelihoods and food security*. IDRC-Earthscan: London.

Robbins, P. T. (2007). The reflexive engineer: Perceptions of integrated development. *Journal of International Development, 19*, 99–110.

Scott, C. A., Faruqui, N. I., & Raschid-Sally, L. (2004). Wastewater use in irrigated agriculture: Management challenges in developing countries. In C. A. Scott, N. I. Faruqui, & L. Raschid-Sally (Eds.). *Wastewater use in irrigated agriculture – confronting the livelihood and environmental realities* (pp. 1–10). UK: CAB International, International Water Management Institute and International Development Research Centre.

Smit, J. & Nasr, J. (1992). Urban agriculture for sustainable cities: Using wastes and idle land and water bodies as resources. *Environment and Urbanization, 4*(2), 141–152.

Tandraatmadja, G., Bum, S., McLaughlin, M., & Biswas, T. (2005). Rethinking urban water systems – revisiting concepts in urban wastewater collection and treatment to ensure infrastructure sustainability. *Water Science and Technology: Water Supply, 5*(2), 145–154.

Toze, S. (2005). Reuse of effluent water – Benefits and risks. *Agricultural Water Management, 80*, 147–159.

Van Lier, J. B., & Huibers, F. P. (2007). *The reversed water chain approach: Optimizing agricultural use of urban wastewater.* 6th IWA Specialist Conference on Wastewater Reclamation and Reuse for Sustainability in Antwerp, Belgium, October 9–12, 2007.

Van Loohuizen, K. (2006). *Afvalwaterzuivering in Nederland: Van beerput tot oxidatiesloot.* RWS-RIZA, The Netherlands: Ministerie van Verkeer en Waterstaat.

Varis, O., Biswas, A. K., Tortajada, C., & Lundqvist, J. (2006). Megacities and water management. *International Journal of Water Resources Development, 22*(2), 377–394.
Warner, J. (2006). More sustainable participation? Multi-stakeholder platforms for integrated catchment management. *International Journal of Water Resources Development, 22*(1), 15–35.
Weber, B., Cornel, P., & Wagner, M. (2007). Semi-centralised supply and treatment systems for (fast growing) urban areas. *Water Science and Technology, 55*(1–2), 349–356.
WHO. (2006). *Wastewater use in agriculture. WHO guidelines for the safe use of wastewater, excreta and greywater* (Vol. 2). Geneva, Switzerland: WHO.

Chapter 13
End User Perspectives on the Transformation of Sanitary Systems

Dries Hegger and Bas van Vliet

Abstract In various Western European countries pilot projects have been set-up in which new waste water management technologies are being experimented in a domestic setting. Domestic end-users often play a crucial role in these projects: ranging from being the main initiators to being the key factor in their collapse. This chapter presents a theoretical appreciation of end-user roles and perspectives in sanitary niche experiments, and develops a toolkit to better understand and experiment with end-user roles and perspectives in new sanitation projects. Subsequently, this theoretical framework is used to analyze two pilot projects in the Netherlands (Sneek and Culemborg). The chapter concludes that an end-user view is instrumental in getting demonstration projects realized as it opens up new ways to link sanitary solutions to end-users' socio-cultural concerns. Furthermore, such an end-user view allows for the successful development and implementation of new sanitation concepts, linking sanitation systems and end-users in various ways.

13.1 Introduction

Several potential technological solutions for more sustainable waste water management systems are available (Lange and Otterpohl 2000; Lens et al. 2001). Past and current pilot projects show, however, that realworld implementation of such solutions is a major challenge. In this chapter we develop a 'toolkit' to understand and experiment with the role of domestic end-users in such pilot projects. In the practice of implementing new sanitation technologies we see that end-users may play various distinctive roles, albeit often accidentally. Several pilot projects have been cancelled because building project managers deemed new sanitation concepts 'unacceptable' for end-users while – on the other hand – in other pilot projects groups of citizens were the main initiator of such concepts.

D. Hegger (✉) and B. van Vliet
Environmental Policy Group, Wageningen University, Hollandseweg 1,
6706 KN, Wageningen, The Netherlands
e-mail: dries.hegger@wur.nl; bas.vanvliet@wur.nl

B. van Vliet et al. (eds.), *Social Perspectives on the Sanitation Challenge*,
DOI 10.1007/978-90-481-3721-3_13,

Our study of pilot sanitation projects in the Netherlands, Sweden and Germany[1] comprises of a review of literature on socio-technical innovation processes and the role of end-users therein combined with a comparative analysis of projects in the three countries. Data collection for the empirical research consisted of desk research, visits to several projects; participatory observation of project team meetings, and semi-structured interviews with project team members and residents in pilot projects.

The next section explores existing literature dealing with innovation processes in socio-technical systems and links it to the Modernized Mixture Approach (MMA, see Oosterveer and Spaargaren, Chapter 2 of this volume). MMA is useful to identifying means to find win–win situations between contemporary waste water infrastructures – often large physical infrastructures that tend to 'lock in' future decision-making or limit innovation trajectories – and sustainable innovations. In the following section we will argue that the MMA approach should be complemented with theories to analyze and understand the socio-cultural concerns of end-users (based on Van Vliet 2006; Shove 2003; Spaargaren et al. 2005; Hegger 2007 Van Vliet and Spaargaren, Chapter 3 of this volume). Then, we illustrate this line of argumentation with empirical findings results of two Dutch pilot projects in Sneek and in Culemborg. The penultimate section discusses the outcomes of evaluating the projects from an end-user perspective, followed by a conclusion and a sketch of the implications of this study for innovation in sanitation.

13.2 Modernized Mixture Approach and Sustainable Sanitary Infrastructures in Western Society

Innovation in sanitation means that win–win situations should be found between the existing large-scale centralized waste water management systems and the various (also) decentralized or decentralizing solutions for sustainability problems that are available nowadays. In line with the contribution of Oosterveer and Spaargaren to this volume (Chapter 2) we use the Modernized Mixture Approach (MMA) as a socio-technical framework to identify solutions for sanitation problems. The Modernized Mixture Approach argues that one should look for 'adequate mixes of scales, technologies, payment systems and cultural and institutional structures that are both economically and environmentally sustainable', in other words a 'Modernized Mixture' (Spaargaren et al. 2005; Hegger 2007), both in developing and developed countries.

The Modernized Mixture approach (MMA) aims to bridge the gap between proponents of 'large is robust and efficient' and 'small is beautiful' that has dominated the debates on infrastructural renewal for a long time. The approach postulates

[1] Conducted within the EET 'Economy Ecology Technology', research program funded by the ministries of Economic Affairs; Education, Culture and Science; and Housing, Spatial Planning and the Environment.

that there is not a single sustainable solution but that there are various solutions, the applicability of which depends on specific temporal-spatial situations. These sustainable solutions are neither completely decentralized systems of the 'small is beautiful' type (Schumacher 1973) disconnecting from modernity altogether. Nor are they an ecologically mindless continuation of large-scale waste water management systems. Modernized Mixtures combine the best of both worlds in several innovative ways (Hegger 2007).

We can define Modernized Mixtures as 'those late modern socio-technical configurations of waste water infrastructures in which various features of simple modern systems have been deliberately and reflexively reconstructed to deal with contemporary social, economic and environmental challenges' (Hegger 2007). For the waste water field a number of relevant variables can be identified to characterize the different socio-technical sanitation systems. Besides classical engineering variables such as the technological scale of waste water management systems and the degree of differentiation of waste water flows, other relevant variables include: 'management scale' (centralized versus decentralized management); 'in use involvement of end-users' (low versus high); 'degree of choice for inhabitants' (low versus high degree of choice) and 'participation of citizen-consumers in the planning phase' (no participation versus full participation). Together, these variables constitute the playground for a transition in waste water systems (ibid). This viewpoint is illustrated in Fig. 13.1.

Figure 13.1 illustrates that Modernized Mixtures aim to transcend and reconcile the traditional opposition between conventional systems and alternatives. By combining the strategic variables in various innovative ways, new concepts for sustainable socio-technical systems can be identified. We need real world experimentation with these sustainable solutions – or 'niches' – to realize them. According to the approach of Strategic Niche Management (Kemp et al. 1998; Hoogma et al. 2002) niches are

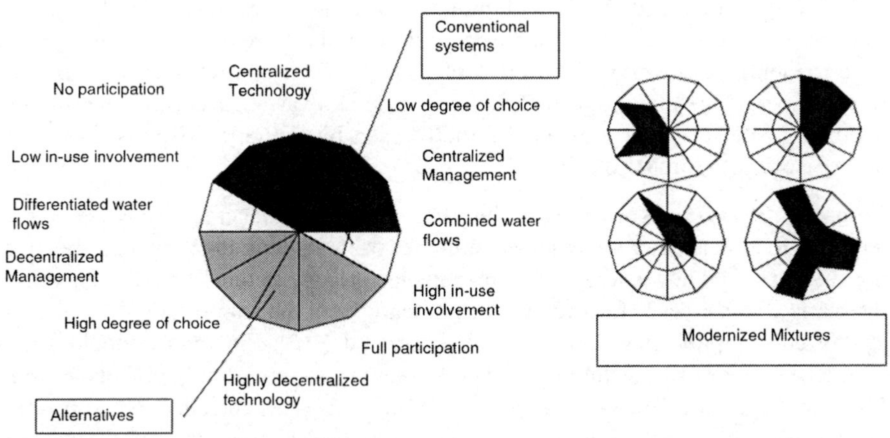

Fig. 13.1 Modernized Mixture approach towards sustainable sanitary infrastructures in OECD countries (Adapted from Hegger 2007)

small-scale experiments in which various actors have the opportunity to learn about the new technologies. They are considered stepping stones for wider changes in socio-technical regimes, thus contributing to sustainable transitions. They aim to protect new technologies from the mainstream market through – for instance – legal exceptions, funding or pilot projects. Thus, niches are thought to provide 'the seeds for change'.

MMA can be used as a 'toolkit' for such niche experiments. The approach enables us to deliberately experiment, not only with various new technologies, but also with different social constellations. This does not mean that Modernized Mixtures would only be feasible at the level of innovative niches only. But the approach does help to arrive at broad experimentation, enabling to find out which socio-technical constellations 'work' and which ones 'do not work'.

13.3 An End-User Perspective on Innovation in Sanitation

Recent literature on the role of end-users in transitions towards sustainability argues that it is crucial to find adequate links between sustainable solutions and end-users' socio-cultural concerns (Hegger 2007; Spaargaren et al. 2007). Such links enable citizen-consumers to see sustainability measures as something that enhances the quality of their consumption practices. Thus, citizen-consumers can become 'change agents': driving forces for sustainable transformations. In line with Brundtland (WCED 1987) such an approach portrays sustainable consumption – and sustainable development in general – not as a loss of quality of life, as something that hurts, but as its opposite.

Linking up with end-users' socio-cultural concerns is more than merely creating acceptance for sustainable solutions or conquering the social and institutional barriers against sustainable transformations. End-users of sustainable solutions should be seen not as barriers but as potential driving forces for transitions towards sustainability (Spaargaren et al. 2007). Different socio-technical systems should be linked towards end-users' socio-cultural concerns in several different ways. To do this, we need to identify relevant variables to look at waste water innovations from an end-user perspective (Hegger 2007). This requires social variables complementing the technologically oriented indicators hitherto used by sanitation experts as well as the variables used in MMA.

Functionality versus social distinction For strategic managers and engineers, it is tempting to look at environmental technologies from a purely functional perspective: how effective and how energy efficient is the technology in removing nutrients from the waste water flows? However, as we can learn from the anthropology and sociology of consumption (e.g. Douglas and Isherwood 1979; Fine and Leopold 1993) products, services and technologies can be seen as socio-cultural symbols as well. Meanings and rationalizations (e.g. radical, normal, [not] for our kind of people) are constructed, contributing to a fit or misfit between sanitary innovations and end-users' life worlds. Thus, next to functionality, social distinction can play an important role. In the most extreme case, this social distinction can take the form of display:

highly conspicuous forms of sustainable consumption (solar panels on the roof, vacuum toilets in the house) can become status symbols, means to show off a sustainable identity towards the outside world (Van Vliet and Stein 2004; Van Vliet and Spaargaren, Chapter 3 of this volume).

Connectedness with nature Through domestic water practices end-users can connect with nature, but also create distance from nature. The wish to connect or to keep distance is often implicit in the actions of end-users. Connecting with nature refers to for instance composting toilets, reed-bed filters or 'Living Machines', whereas distance from nature is reflected in the use of chlorine to clean toilets, or in waste water management systems that have been designed as high-tech systems disguising the link with nature.

Dependencies between end-users. The behavior of one end-user can have more or less direct influence on others. Waste water innovations often lead to more direct dependencies between households (Van Vliet et al. 2005). 'Misbehavior' can affect other households in the same neighborhood. In this respect we can think of the use of vulnerable treatment systems (reed bed filters) used by several households. Here the social interdependencies are more direct when compared to central sewer systems. Sanitary facilities shared by several persons within the same household illustrate another very common form of interdependency: for instance the responsibility of keeping them clean and tidy.

In-use involvement. In-use involvement refers to whether people (have to) take up extra tasks in the use and management of sanitation systems. From the perspective of engineers and project managers the question is whether end-users should be given extra responsibilities if compared to conventional systems/situations. From an end-user perspective the question is how household members divide tasks; whether they can fit these extra tasks into their daily chores and whether they have the necessary skills to carry them out.

Power of end-users vis-à-vis systems of provision. Environmental innovations differ in the degree to which they help empower citizen-consumers. In case of grid-connected systems, implemented in a top-down fashion, the power of individuals to influence these systems is relatively low. Theoretically, this power is higher in case of stand-alone devices for which end-users can choose individually. The latter path is relatively under-explored (with some exceptions though) in the waste water field, where the idea of uniform provision (each household the same system) is deeply rooted (Hegger 2007). The degree of citizen-participation (e.g. when setting up a new neighborhood) also influences the power of end-users.

The five strategic variables discussed above are crucial dimensions of end-users' socio-cultural concerns. Looking at these concerns enables us to find out whether, to what extent and in what ways sanitation innovations can be geared towards these concerns. Figure 13.2 illustrates that we can also distinguish conventional systems, alternatives and Modernized Mixtures from an end-user perspective.

It should be acknowledged that the needs and demands of end-users cannot be taken for granted. Rather, 'knowledge' about these needs and demands should be seen as assumptions which should be checked and challenged. By checking such

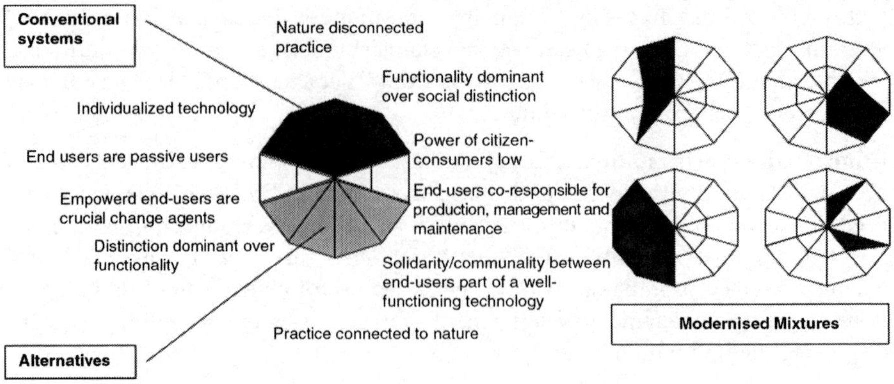

Fig. 13.2 Modernized Mixtures approach from an end-user perspective (Adapted from Hegger 2007)

assumptions we can anticipate on the use of socio-technical innovations. If we can find means to acquire such knowledge before large-scale implementation of new technologies in demonstration projects, this likely prevents practical problems and end-user resistances and may enhance learning experiences in pilot projects.

An end-user perspective enables project executors to look for sustainable solutions from a systemic perspective (cost effectiveness, eco-efficiency, technical scale etc.) and – at the same time – from a life-world perspective. End-users can play various roles in sanitation pilot projects, for instance as initiators, participants, or as respondents in consumer research. The following section discusses two cases of innovation in sanitation from such a combined systemic and life-world perspective.

13.4 Innovation in Sanitation: Dutch Pilot Projects Sneek and Culemborg

Most demonstration projects experimenting with radical innovations in sanitation are of a relatively recent date (from 1990s on, although the first scattered experiments took off in the 1970s). These experiments can often be seen as the successors of experiments with less path-breaking water- and sanitation-related innovations in the home, such as water-saving taps and showerheads or disconnection of rainwater from the sewage system.

We argue, following Verheul and Vergragt (1995), that a distinction can be made between expert-led and citizen-consumer driven experiments (Hegger 2007). The former concern projects carried out by institutional actors such as engineering departments, firms and governments whereas the latter are set-up by citizens-groups and (eco)-NGOs, generally committed to sustainable development and therefore willing to invest much time and money to get pilot projects realized. The current section reviews one expert led experiment (Lemmerweg-Oost in Sneek) and one citizen-consumer driven experiment (Eva Lanxmeer in Culemborg). Both cases concern new building areas in medium-sized Dutch municipalities.

13.4.1 *Successful Expert-Led Innovation: Lemmerweg-Oost – Sneek*

The Lemmerweg-Oost project in Sneek comprises a new building area consisting of 32 rental apartments. At the end of 2005 a consortium was formed in which the municipality and two housing corporations collaborated with a group of environmental engineers, the regional Water Board and a company specialized in sewage and waste water treatment to realize a pilot project. The technological system entails the separate collection of the black waste water from the toilets in the neighborhood using vacuum pipes. The collected waste is treated in an anaerobic digester located in the neighborhood.

The project in Sneek was set up after two earlier efforts to set up a similar demonstration project in other municipalities failed.[2] In these projects, employees of the involved municipalities and project developers were skeptical about the proposed technological system. Amongst others, they deemed the sanitary solutions 'unacceptable' for end-users because of smell- and noise-related concerns. Furthermore, in both failed projects the question 'how to manage the implemented systems after ending the demonstration phase of the project' (2 years, until the end of the EET-DESAR program) was put on the table.[3] In Sneek the involved actors were more favorable towards setting up a demonstration project. A crucial reason for this was that in Sneek the proposed environmental innovations linked up with the interest of the municipality and the housing corporation. The former made the 'vacuum toilets' project part of its broader PR strategy to position Sneek as a 'Water City'. The latter could link the demonstration project to its corporate social responsibility strategy.

The first residents moved in their new houses and started using vacuum toilets in June 2006. In October 2006, 16 residents from different households were interviewed on their first experiences with the new technologies. It turned out that the majority of the residents favored the new sanitary system, notwithstanding the fact that all interviewees reported nuisance from the noise the toilet systems produced. These unanticipated noise problems considerably influenced the perceived quality of living in the neighborhood. The installation of silencers in all houses some time after the interviews reduced this nuisance considerably.

[2] The pilot project was set up in the framework of the EET-DESAR project. EET is a Dutch abbreviation which stands for 'Economy Ecology Technology'. It was a research program funded by the Ministries of Economic Affairs; Education, Culture and Science; and Housing, Spatial Planning and the Environment and executed by SenterNovem. DESAR stands for 'Decentralized Sanitation and Reuse'.

[3] The project partners in Sneek deliberately chose for a management structure similar to the situation where conventional waste water management systems are applied. That is: the municipality is responsible for the 'conventional' sewage pipes used for the collection of grey water in the neighborhood; the Waterboard for waste water treatment (but not of the black fraction of the waste, which was managed by the company installing the anaerobic digester). See also: Hegger (2007) (on Wageningen and Emmen), Van Vliet and Stein (2004) (Wageningen).

The project in Sneek can be seen as a top-down demonstration project for innovation in sanitation. The domestic end-users have been involved, both in the planning phase and in the actual use phase of the project. Based on these end-users perceptions and experiences, adaptations such as the installation of the silencers have been made, illustrating that top-down innovation can be combined with an end-user perspective.

13.4.2 Successful Citizen-Consumer Driven Innovation: EVA-Lanxmeer – Culemborg

The EVA-Lanxmeer project in Culemborg includes a new residential area, office buildings and an 'EVA-Centre' in which the principles behind the project are made visible to large (visitors-) groups, to inform Dutch society. The project uses sustainable energy and mobility concepts in combination with ideas about social sustainability (e.g. combination of rental and private houses; urban design promoting social interaction). The residential area consists of approximately 200 rental and private houses, the building of which began in 1999. The initiators of this new building area consisted of a group of citizens, who envisaged a new sustainable residential area 'somewhere in The Netherlands' and subsequently involved the municipality of Culemborg in these plans.

Sustainable urban water- and waste water management has been applied as one out of several social and environmental sustainability innovations in this pilot project (Hegger 2007). Rain water, grey water (water from kitchen sinks, shower and washing machine) and black water (waste water from the toilet) are separately collected and treated. Grey water is led to a separate sewer system (not used for black water and rainwater) and is treated in reed-bed filters. The effluent from these filters is transported from the neighborhood via the same ditches that remove rainwater. Black water is collected with a third sewage system. The toilets applied in the area are so-called booster toilets (Gustavsberg toilets) consuming a relatively low amount of water (2.5–4 l per flush).

The residents in the neighborhood display a high degree of environmental consciousness. Interviews carried out in 2003 (Hegger 2007) indicate that the residents generally believed that 'there are some things you have to do or leave if you want to live in this neighborhood'. A large number of respondents indicated that the sustainable character of the neighborhood they live in forms an important part of how they see themselves and how they construct their lifestyles and personal identities.

13.4.3 Innovation in Sanitation: A Systemic Perspective

Figure 13.3 depicts the two projects when perceived from a systemic perspective. These two projects, although different in their set-up, can be called successful as the pilot projects have been actually realized and working technological

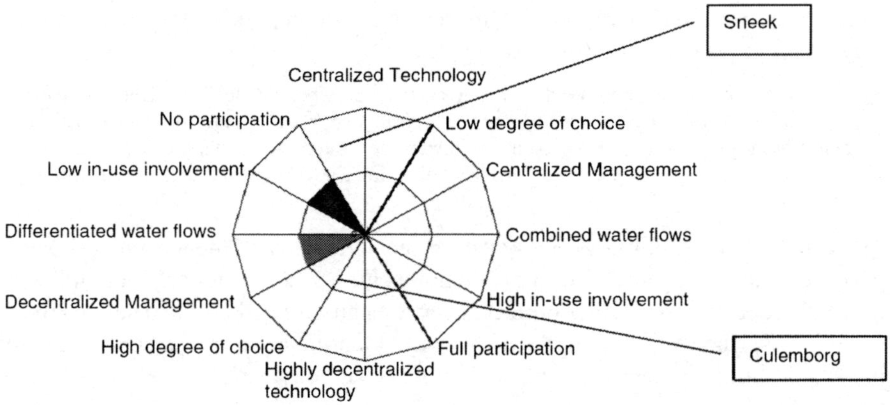

Fig. 13.3 Sneek and Culemborg approached from a systemic perspective (Adapted from Hegger 2007)

innovations are in place. In both projects field research among end-users was considered necessary, illustrating that in both projects (in Sneek based on the lessons of past failures) the need for an end-user perspective has been acknowledged.

13.5 Analyzing Sneek and Culemborg from an End-User Perspective

The projects in Sneek and Culemborg differ considerably, with regard to the implemented socio-technical systems, the way in which the projects were organized, the way in which the systems have been implemented as well as with regard to the socio-cultural concerns of end-users living in both neighborhoods. Nevertheless, in both projects a 'working link' between providers and end-users could be established: sanitary innovations were actually applied and used in practice. In this section we will discuss both projects from an end-user perspective.[4]

Functionality versus social distinction. The residents in both pilot projects are very well-aware that they are living in a demonstration project. In Sneek, some residents had the feeling that they 'are different' because their toilet system is 'new', 'modern' and 'the first in The Netherlands'. The residents also indicated that they were aware of the frequent guided tours in their neighborhood. In Culemborg the aspect of social distinction is even more outspoken. Residents indicate that they

[4] For both cases project executors (three in Sneek, five in Culemborg) and end-users (18 in Sneek [October 2006], 15 in Culemborg [March 2003]) have been interviewed. Other data collection methods include desk research, participatory observation of project team meetings and site visits.

clearly see their neighborhood as 'different from the rest of Culemborg' and that they 'want to provide a positive example'.

> Positive about this neighborhood is the 'it's our type of people' feeling. Everyone feels responsible for the environment. Where we used to live, people found it strange that we didn't have a car, here it is accepted. We always separated our waste, were careful with water and energy use. That's much easier here. (Interview with one of the residents in Culemborg)

The residents in Culemborg clearly use the eco character of their neighborhood to play the distinction game. Social distinction is higher in Culemborg than in Sneek, especially because the former project is carried and initiated by the residents themselves. Nevertheless, also in Sneek we see identification of residents with the installed sanitary innovations, which they frame as 'modern':

> We are the first area in The Netherlands where they have this system. (Interview with one of the residents in Sneek)

These examples illustrate that – contrary to the common sense expectation that sanitation is, and should be, highly inconspicuous – sanitary systems can have a display dimension (see also Van Vliet and Spaargaren, Chapter 3 in this volume). In both projects we see some degree of active display by providers as well as consumers. This has potential positive and negative consequences from an environmental sustainability point of view. On the one hand, social distinction through sanitary innovations can become an extra asset in quality of living and a means to link such innovations to end-users' concerns. On the other hand, there is the risk of creating eco-gated communities – as the EVA-Lanxmeer project is perceived by some outsiders – complicating the dissemination of the innovations to broader groups of end-users.

Connectedness with nature. The innovations in both projects link end-user behavior more directly to the functionality of waste water treatment processes. For instance, in both projects the use of chlorine-containing detergents is a risk for waste water management systems. In Culemborg this 'nature connectedness' was put forward as a positive aspect. The sanitation systems in the neighborhood have been integrated in a more encompassing concept in which amongst others water management and enhancement of biodiversity play a role. As one of the residents in the project in Culemborg indicates, the water system also has an important control function: because the reed bed filters are sensitive, people are obliged to be careful with it. If it would be removed, the urge to be environmentally friendly would diminish, and so would be the feeling of 'we are in it together'.

The project executors in Sneek, however, framed nature connectedness as something negative, as a (public health and aesthetics-related) barrier from an end-users' point of view. This maybe explains why in the latter project, the design of the waste water management system (a robust looking vacuum system) more or less 'disguises' the 'nature connectedness' of the waste water management system. Interviews with consumers in Sneek, however, indicate that the residents are generally well-aware that the innovations in their neighborhood are intentional environmental innovations.

The majority of the residents indicated that they were aware of the environmental end-goals of the project (decreased water consumption, biogas production, reuse of nutrients and more efficient waste water treatment), an indication for a successful information provision by the project team.

Dependencies between end-users. In Culemborg interdependencies between end-users were 'deliberately' created. The more direct link between end-users (one resident using chlorine affects the quality of living of many others) can be seen as an integral part of initiators' vision in which closer social ties between end-users are seen as something positive. Interviews with the residents indicate that most of them regularly go to meetings. Most visited are the meetings of their 'hof', a group of houses grouped together in such a way that social interaction is promoted. The 'hof' (and not so much the neighborhood at large) seems to be the level at which social cohesion and identification is highest. For one of the interviewees, an architect by profession, this is even a reason to argue that the water management system within the neighborhood should have a technical scale at 'hof' level. It would be more efficient and people would be more involved with the system if it would be literally on their doorstep.

In this respect EVA-Lanxmeer differs from the project in Sneek, where dependencies between end-users were created more 'accidentally'. Another characteristic of the project in Sneek is that the project leader himself lives in the neighborhood. The fact that 'he has the same problem as we have' and that it is 'in principle possible to approach him with problems' has led to trust of the residents in the technologies and in the experts behind the technology.

Both examples show that social interdependencies can be created in different ways. Whether such interdependencies are preferred or not depends on the specific social, temporal and spatial circumstances in a pilot project. It is worthwhile to deliberately reflect on the desirability and precise form of these social interdependencies rather than to just let them emerge. This reflection should include both the planning and building phase as well as the actual use phase of projects. For instance, who will take over the responsibilities if the project leader moves out of the Sneek project? And what will happen with all formally and informally created interdependencies in Culemborg if a first generation of eco-committed residents would be succeeded by a less committed second generation?

In-use involvement. In Sneek the in-use involvement of end-users is limited to some conscious awareness with regard to the use of the sanitation innovations. The situation in Culemborg is equal to the majority of end-users, although some end-users take up extra tasks in task-forces. One of these task-forces also carries out some management and maintenance tasks for the reed-bed filters. Both projects thus comprise some extra in-use involvement compared to conventional systems although this in-use involvement is relatively limited compared to the in-use involvement related to the management and maintenance of, for example, composting toilets (Krantz 2005). For instance, the project team in Culemborg aimed to set up systems in which residents do not have to compromise the level of comfort they are accustomed to:

> And actually I think that it has to be organized in such a way that an average inhabitant of a neighborhood does not notice whether he is connected to a conventional waste water system (…) or lives in a neighborhood like this (…) in principle I think that you should not have to compromise your level of comfort, in order to make it succeed. (Interview with a member of the project bureau in Culemborg)

In-use involvement seems to be most important in the planning phase and at the start of the actual use phase, when the end-users are in the midst of a process of de- and re-routinization. At these moments, the residents are in a process of adapting their domestic (waste)water practices, while at the same time the probability of problems is highest:

> We installed for example the so-called boosters and in practice this confronted us with a lot of troubles, because they were clogged and then you had to (…) in your house (…) in the hall there was such a thing which had to be opened and it smells like hell of course (…) it did not lead to resistance of the inhabitants, so a high degree of tolerance was present (…) also we did not have problems with the installation of the reed-bed filters, which had the reputation that they could smell in winter, but no one made a big problem of that. (Interview with one of the residents in the project in Culemborg, member of the 'taskforce energy and installations' in the neighborhood)

Power vis-à-vis systems of provision. In Sneek the power of the end-users vis-à-vis the providers of the sanitary innovations was low. In fact they were not given the choice of being in favor of or against the sanitary system as implemented. This system was part of a 'package deal' (the house) which they could accept or decline (the latter of course only if the residents had acceptable alternatives). However, it seems that many residents in Sneek found it reassuring that their housing corporation participated in a demonstration project and this bestowed the innovation with extra legitimacy in the eyes of the end users. Also, the residents seemed to be sympathetic towards the environmental end-goals of the project. As one of the residents put it: 'well at least they have to experiment somewhere'. Also the fact that the project leader lives in the neighborhood contributed to the legitimacy of the project in the eyes of the residents.

In Culemborg end-users' power was relatively high in the planning phase of the project, when the first group of residents mobilized other social actors. End-users were a crucial force for getting the project realized. Such power is, however, much lower for newer residents. Similarly to Sneek, for these residents the systems were part of a package deal: there was no other choice than to accept the systems that were already there.

13.6 Conclusions

Two demonstration projects preceding the project in Sneek were cancelled in their planning stage due to normative claims about the future end-users (Hegger 2007). For this chapter, we selected two successful pilot projects, in Sneek and Culemborg, which were actually realized. As we have shown, each of these projects – sometimes deliberately, sometimes accidentally – succeeded to fit sanitation innovations (and

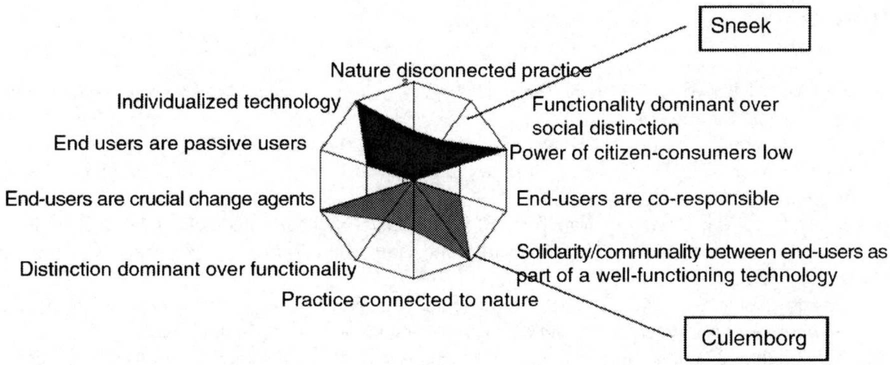

Fig. 13.4 Sneek and Culemborg assessed from the variables of an end-user perspective

other environmental innovations) to end-users' socio-cultural concerns. As Fig. 13.4 illustrates, this fit was brought about in different ways.

This chapter draws upon the Modernized Mixture Approach (MMA) as a theoretical framework to analyze the involvement of end-users in sanitary pilot projects. The MMA should not restrict itself to so-called systemic variables; also the involvement of domestic end-users should be specified. We have denominated five strategic variables which can be used for this purpose: (1) functionality versus social distinction; (2) connectedness with nature; (3) dependencies between end-users; (4) in-use involvement; (5) power vis-à-vis systems of provision.

The five end-user variables can be used to extend the toolbox for transition experiments: niche experiments are not only designed from a systemic perspective but also from a life-world perspective. This is to be preferred above allowing life-world dimensions just emerge into niche projects, something often happening in practice. A next step would be to set up new strategic niche experiments which deliberately depart from these end-user dimensions (Hegger et al. 2007). The results of such experiments can be used to specify and refine theoretical insights on end-user perspectives as well as to derive practical lessons about the tested systems, resulting in learning for sustainability.

Waste water infrastructures are loosing their uniformity and taken-for-grantedness, just like other network-bound systems such as the electricity grid did in the past. We can expect a future in which waste water infrastructures and end-users are linked to one another in different ways. Different Modernized Mixtures – from a systemic *and* from an end-user perspective – imply different social, temporal and spatial-specific socio-technical constellations which have to 'prove' their relevance and added value over and over again. The current dominant sewer based sanitation system will most likely be replaced not with *one* but with *many* innovative (and hopefully sustainable) systems linking to end-users, socio-cultural concerns in several ways.

References

Douglas, M. & Isherwood, B. (1979). *The world of goods – towards an anthropology of consumption*. London: Allen Lane.

Fine, B. & Leopold, E. (1993). *The world of consumption*. London: Routledge.

Hegger, D. L. T. (2007). *Greening sanitary systems: An end-user perspective*. Ph.D. thesis, Wageningen University, Wageningen.

Hegger, D. L. T., Van Vliet, J., & Van Vliet, B. J. M. (2007). Niche management and its contribution to regime change: the case of innovation in sanitation. *Technology Analysis & Strategic Management, 19*(6), 729–746.

Hoogma, R., Kemp, R., Schot, J. W., & Truffer, B. (2002). *Experimenting for sustainable transport: the approach of strategic niche management*. London/New York: Spoon Press.

Kemp, R., Schot, J. W., & Hoogma, R. (1998). Regime shifts to sustainability through processes of niche formation: the approach of strategic niche management. *Technology Analysis and Strategic Management, 10*, 175.

Krantz, H. (2005). *Matter that matters: a study of household routines in a process of changing water and sanitation arrangements*. Ph.D. thesis. Linköping University, Linköping.

Lange, J. & Otterpohl, R. (2000). *Abwasser, Handbuch zu einer zukunftfähigen Wasserwirtschaft*. Donaueschingen: Mall-Beton-Verlag.

Lens, P., Zeeman, G., & Lettinga, G. (Eds.). (2001). *Decentralized sanitation and reuse: concepts, systems and implementation*. London: IWA Publishing.

Schumacher, E. F. (1973). *Small is beautiful – economics as if people mattered*. London: Blond & Briggs.

Shove, E. (2003). *Comfort, cleanliness and convenience: The social organization of normality*. London: Berg.

Spaargaren, G., Mommaas, H., Van Den Burg, S., Maas, L., Drissen, E., Dagevos, H., et al. (2007). *More sustainable lifestyles and consumption patterns: a theoretical perspective for the analysis of transition processes within consumption domains*. Contrast Research Report, TMP project. Environmental Policy Group, Wageningen University; Telos: Tilburg University; Bilthoven: RIVM; The Hague: Lei.

Spaargaren, G., Oosterveer, P., Van Buuren, J., & Mol, A. P. J. (2005). *Mixed Modernities: towards viable urban environmental infrastructure development in East Africa*. Wageningen: Wageningen University.

Van Vliet, B. J. M. (2006). The sustainable transformation of sanitation. In J. P. Voss, D. Bauknecht & R. Kemp (Eds.), *Reflexive governance for sustainable development*. Cheltenham: Edward Elgar.

Van Vliet, B., Chappells, H., & Shove, E. (2005). *Infrastructures of consumption: environmental innovation in the utility industries*. London: Earthscan.

Van Vliet, B. J. M. & Stein, N. (2004). New consumer roles in wastewater management. *Local Environment, 9*, 353.

Verheul, H. & Vergragt, P. J. (1995). Social experiments in the development of environmental technology: a bottom-up perspective. *Technology Analysis and Strategic Management, 7*, 315.

World Commission on Environment and Development. (1987). *Our common future*. Oxford: Oxford University Press.

Chapter 14
Conclusion and Discussion

Gert Spaargaren, Bas van Vliet and Peter Oosterveer

Abstract In this concluding chapter, the aims that were formulated at the start of this volume are revisited. Drawing on the specific social scientific way of defining and trying to meet the sanitation challenges, the most important results of this volume are presented. Building upon the results achieved, the future of the field of sanitation as an area of social scientific research and policy making is explored.

14.1 The Sanitation Challenge

According to the Brundtland report (WCED 1987) the concept of sustainable development combines economic development with an ecological and a social dimension. The 'Sanitation Challenge' addressed in this book can be said to refer to the key dimensions of sustainable development.

Given the context of the highly developed parts of the world, the long term 'ecological sustainability' of existing sanitation infrastructures is the most prominent concern. Sustainability in this case means first and foremost ecological sustainability and the challenge consists of constructing sanitation systems which use less water, are climate proof, and deal with nutrients in a more sustainable way. Building upon the experiences of large numbers of pilot projects in newly build areas in particular, the sanitation challenge can be met when the more sustainable experimental systems for dealing with waste-water and (human) wastes are up-scaled to the level of the existing housing stocks and existing large scale infrastructures. When developing niche-experiments into mainstream processes of design and decision-making about urban environmental infrastructures, the sanitation challenge in this context should result in a new sanitation 'regime'. A regime which better fits the challenges of the time by combining a high level of service with an improved environmental performance.

G. Spaargaren (✉), B. van Vliet, and P. Oosterveer
Environmental Policy Group, Wageningen University, Hollandseweg 1, 6706 KN, Wageningen, The Netherlands
e-mail: gert.spaargaren@wur.nl; bas.vanvliet@wur.nl; peter.oosterveer@wur.nl

B. van Vliet et al. (eds.), *Social Perspectives on the Sanitation Challenge*,
DOI 10.1007/978-90-481-3721-3_14, © Springer Science+Business Media B.V. 2010

In the context of the developing parts of the world, the sanitation challenge can be said to refer primarily to the social dimension of sustainable development as articulated in the Brundtland report and in the Millennium Development Goals (MDGs) in particular. Although environmental or ecological issues are not put aside, they are always discussed in direct relation to issues of human health and accessibility. Since the 'Hygienists' movement in nineteenth century Europe (Van Zon 1986) it has been acknowledged that sanitation challenges are about preventing diseases and improving the hygienic quality of life especially for the urban poor. Targeting the fast growing number of people in unplanned settlements and urban slums, urban environmental infrastructures are judged primarily with respect to their accessibility in social, physical and above all economic terms.

The Brundtland philosophy of sustainable development has been operationalized into concrete aims and targets in the Millennium Development Goals. This UN-guided process has gained global support. Almost all chapters in this volume use the MDGs as the accepted way of framing the sanitation challenge. The MDG-setting process is recognized for its clear targets within a fixed time period. In this way, the urgency of the challenge is emphasized, and the plea for acceleration in the process of modernizing urban sanitation infrastructures is given full priority. Although MDGs are clear and fixed, the route for arriving at the designed destination is not outlined in much detail. Especially when talking about sanitation, the best way forward – the basic path of modernization to be chosen – is open for discussion. For some, the 'simple modernization' model as represented by the sanitation revolution in nineteenth century Europe, is still the leading example, to be copied in the developing world as frequently as possible. For others, the negative, environmental side-effects of the old modernization model prove the need for a different approach, labeled 'reflexive modernization'.

This volume aims to contribute to the realization of MDGs by discussing the best possible ways to design the process towards the goals. Instead of focusing on the political feasibility of the goals to become realized (or not) within the fixed limits of time, the group of authors prefer to reflect about the modernization path to be chosen and the kind of urban sanitation models best suited to realize the MDGs. As indicated in the introduction to this volume, three specific contributions were high on our agenda when designing it. First, a discussion and elaboration of the theory of the Modernized Mixtures aimed to offer a third-way approach in between the centralized or radically decentralized options and positions as they were so characteristic for the sanitation debate in the previous millennium. The Modernized Mixture approach has as its leading principle that physical, spatial, and social conditions 'on the ground' have to be taken as starting point for designing, discussing and implementing sanitation facilities and infrastructures which are assessed with respect to the best 'performance mix' of both ecological and social elements. Second, because the sanitation solutions are no longer 'one fit for all' but instead are made context-dependent and locally specific, tools need to be developed to design the optimal mixture for each situation. For that reason, 'decision support systems' gain specific importance. A number of decision support systems are offered in this volume, different in several respects but all emphasizing the need to combine technical and social design criteria. The inclusion of social criteria in the design, construction and maintenance of sanitation systems is relatively new to the field of sanitation and more difficult to realize than most scientists, policy makers and practitioners had expected. So the third major goal we outlined for the book project is a thorough exploration of the social dimension of the sanitation challenge. We explored what exactly is meant with 'social', and how an emphasis on social dynamics can be combined with the prevailing technological rationality that is dominating the field of sanitation.

These were the themes and priorities we had in mind when designing the book. From the onset it was clear that discussing the sanitation challenge meant something different in the North when compared to the South and we decided that both perspectives should be represented in the volume. And although being aware of the difficulties that it would raise, we were convinced about the need for an open dialogue and more serious forms of collaboration between technical engineers and social scientists. So the contributions came from teams of authors, sometimes social scientists, sometimes engineers, and in some cases a mixture of both. It is against the background of themes and priorities that we shortly want to make up the balance of the book.

14.2 The Balance of the Book

In the first part of the book, the concept of Modernized Mixtures is used to refer to a theoretical perspective we think is innovative and in particular relevant for the present stage in the debate on the provision of sanitation services. Although in this volume the theory of Modernized Mixtures is used to analyze waste water and waste provision in particular, this perspective is not limited to any particular set of material flows. The theory is developed for analyzing the provision of utility services to households. It pertains to the specific parts of domestic consumption which are made possible by providers who organize their service delivery with the help of socio-technical networks. Through these networks, the products and material flows are 'delivered' at the level of the households under specified conditions and with the use of material infrastructures like pipes, cables or frequencies. For this reason, the Modernized Mixtures approach can be said to refer to the analysis of the 'infrastructures of consumption' (Van Vliet et al. 2005) in general. As Oosterveer and Spaargaren argue in their contribution to part I, the social and technical dimensions of the networks and infrastructures of provision has to be analyzed in close connection. Just rolling out material infrastructures without knowledge of the specific local social and political conditions and without taking into consideration the culture, the habits and the buying power of the supposed end-users of the services, will result in failed service provision. They in particular set themselves the task of specifying the questions that social scientists should answer when analyzing infrastructures of consumption. Using sanitation and waste provision in East Africa as their case study, they open up the black box of behavior, politics and culture by identifying the key actors and dynamics that determine success or failure of service delivery at the level of households, neighborhoods, municipalities/cities, and nations. They show that the so called lock-in effects of service provision do not primarily refer to the technological dimension of service provision but instead are rooted mostly in the socio-political and cultural landscapes of East Africa. When people move from the remote rural areas with very low population densities into urban slums with very high population densities, they are forced to adapt also their sanitary behaviors to these new conditions.

As Van Vliet and Spaargaren in their chapter to part I indicate, behavioral routines which are deeply rooted in cultural traditions can serve as important constraining factors when designing new sanitation infrastructures. When a substantial part of the population is Muslim, the introduction of dry toilet systems could be problematic since the user practices which come along with those systems run counter to some religious prescriptions. This constraining power of culture is not limited to the LDCs, as the authors make clear. In many OECD countries, end-users don't like to 'sensory experience' waste-flows of any kind in a direct way. They show that the paradigm shift of making the infrastructures of provision for energy and water more visible, transparent, accessible and enjoyable for end-users seems to halt before (sanitation) waste flows. People want to 'flush-and-forget' as soon and effectively as possible and do not seem to be interested in making the processes of delivery, transport, treatment and reuse of (sanitation) waste flows accessible or sensible for end-users. As a result, they argue that the cultural analysis of waste-flows should be distinguished from water and energy flows in some crucial respects. This is not to say that re-introducing the senses as an outcome of innovation in sanitation is impossible, rather it requires great care for the specific socio-cultural beliefs and practices around sanitation. With their analysis Van Vliet and Spaargaren also illustrate why in general the provision of water or energy is considered politically 'hot' and more popular among donor agencies when compared to sanitation infrastructures which are just considered 'necessary' and 'important' most of the times.

Part I is concluded with a chapter by Okot-Okumu and Oosterveer on the sanitation situation in Uganda. This chapter more or less represents a type case of (the lack of) sanitation infrastructure provision in many LDCs. Substantial parts of the population do not have access to proper sanitation infrastructures, and especially in poor neighborhoods the living conditions and life expectancy can be shown to be negatively influenced by this lack of sanitation facilities. They make an inventory of the type of sanitation facilities which are available in urban centres in Uganda, and they specify the conditions which make sanitation facilities (un)safe. While in principle the decentralization of politics in Uganda offers new perspectives and the legal and political institutions for a more effective development of sanitation infrastructures, in practice the existing institutions 'do not deliver'. The overall picture is one of promises and potentials for the next future, while for the present situation 'failed provision' is the most common situation. It is against this background that the authors make a case for a realistic, locally adapted approach of infrastructural provision, in correspondence with the theory of the Modernized Mixtures.

In the second part of the book the focus is on the operationalization of some of the general principles of the Modernized Mixtures Approach. When designing new systems, making decisions on existing and new infrastructure and implementing them, a vast range of criteria and factors have to be taken into consideration. These criteria have to be known to and used by engineers, policy makers and end-users at specific moments in the process. Because they are tools used in decision-making, they are labeled as 'decision support systems' and they take the form of 'Multi-Criteria-Decision-Analyses' (MCDAs) most of the times. Depending on the phase of the decision-making process and the key actors involved in the decisions, the specific focus of the decision-support-systems may vary. The chapters in part II of this volume are all dealing with MCDAs and decision-support-systems, each highlighting different elements in particular.

The chapter by Tilley, Zurbrügg and Lüthi shows great ambition when setting the task of developing a 'common language for the technical communication about sanitation'. This common language is build around concepts like products, flow-streams, technologies, processes and systems. It has its origin in system theory and is detailed and precise in character. The fact that this approach stems from a large international group of researchers involved in an EU-research project gives support to their claim of making this language into a real transnational phenomenon, to be used both in situations of over- and underdevelopment of sanitation infrastructures. As the authors themselves are the first to admit, this common language is focused on the technical dimensions of sanitation provision and should be used primarily to facilitate the communication between experts and researchers, rather than with practitioners and end-users.

In their contribution, Van Buuren and Hendriksen make this structured dialogue with practitioners and end-users the core concern of their MCDA. To facilitate the dialogue about sanitation systems with non-experts, they designed a 'language of tokens' describing toilets, kitchen-sinks, baths and showers and the multitude of sanitation configurations they make possible. The more than 50 options for urban sanitation in developing countries are screened, selected and discussed in a series of stakeholder workshops the authors organized in different parts of the world. In doing so, the perspective of the end-users and non-paid professionals is included in the process of decision-making about sanitation infrastructures. What results is basically *politizing* the process of decision-making, with choices made explicit, discussed among stakeholders and decided upon with the involvement of all major parties. While recognizing that in most cases the existing power relations do not allow for a completely open, bottom-up process of decision-making, the authors make a plea for the inclusion of the non-expert and non-professional in order to optimize the sanitation solutions decided upon.

The chapter by McConville, Kain and Kvarnström is in some respects close to the previous chapter of Van Buuren and Hendriksen. They also discuss decision support systems from the perspective of local stakeholders, arguing that their knowledge and preferences are important in the design, operation and management of sanitation systems. Analyzing the literature on this point, they compare the criteria emphasized by technical experts with the criteria put forward by local actors and institutions working at the local levels. As the main results from their case studies in West Africa they report local actors not to be much concerned about technological performance of sanitation systems or about issues of health and (ecological) sustainability. Instead, local actors emphasize the long term functioning of systems, issues of participation and the need for an embedding of sanitation systems in local culture and politics.

In their contribution to part II, Meinzinger, Ziedorn and Peters also reflect on decision-making about possible system configurations, this time in the context of societies with already highly developed sanitation infrastructures. Using Germany as their case-study, they show how a reflexive approach to sanitation for Germany means the inclusion of both the social and spatial/physical initial conditions. They use the example of urine separation in retro-fit situations of existing housing stocks to discuss the key factors which should be taken into consideration during the decision-making on the best possible technical solutions. The costs for the construction and operation of urine separation systems depend very much on the density of the population, the spatial structure and the types of houses in a certain area. Sometimes individual urine collection tanks are to be preferred, while in other cases shared facilities are the best solution. From their analysis of ecologically advanced sanitation in highly developed societies it becomes clear as well that the introduction of source-separating sanitation systems brings along drastic changes for end-users. They not only have to drastically adapt their sanitation practices of showering and toilet use, they also have to make extra space available for new equipment which brings along significant extra costs. It might take some good arguments indeed to convince householders about the need of preventing nutrient losses when they are loosing 1 m^3 of space and 2,000 euros for the more sustainable sanitation solution offered.

In their chapter on decision-making for urban sanitation infrastructures in East Africa, Letema, Van Vliet and Van Lier are faced with starting conditions that are very different from the German situation. Sanitation infrastructures are absent or at best fragmented, and socio-spatial conditions do not allow for uniform (centralized) solutions. Against this background, decision-making on sanitation systems relevant for the local situations can best be guided by a Modernized Mixtures Approach, so the authors argue. They suggest using the concepts of networks (material) flows, and spaces to assess the initial situations in theoretical terms. When making decisions on the best possible solutions, also the (bad) performance of existing sanitation systems have to be taken into account. How can the existing structure be improved, and how can one build upon the existing fragments in order to combine them into sanitation networks of greater geographical scale and technical scope in the future? The authors use the cases of Kampala (Uganda) and Kisumu (Kenya) to discuss the planning and decision-making for improved sanitation infrastructures in the next and distant (2030) future. Using existing information on (projected) densities and (sanitation) flows, they show that different technical configurations tend to result for different local spaces. In doing so, the authors show the policy relevance of the Modernized Mixtures approach in a convincing way.

The chapter by Toubkiss which concludes part II is about one of the most crucial non-technical variables in sanitation provision: money. Although recognized as a pre-condition for success by all authors, hardly any in-depth studies on financing sanitation of the post-centralized system kind has been published so far. Most economic studies available show a bias towards big systems while completely neglecting on-site options and semi-collective solutions adapted to local spaces. Toubkiss develops an economic analysis for locally adapted sanitation configurations, using Sub-Saharan Africa as his case. When the big money of international donors does not fit to local conditions and the local authorities can be shown to be almost completely absent in sanitation projects at the de-central level, what are the feasible ways of financing local sanitation infrastructures? "The main problem in financing sanitation infrastructures is to be found at the local level", so Toubkiss argues. He explores the potential of micro-credit schemes for doing this job, but concludes that in case of sanitation these local arrangements run the risk of putting poor households (further) into debt. The best financial solution seems to be a targeted, circumscribed subsidy on sanitation facilities. Also ways of 'indirect' financing of sanitation facilities and infrastructures (along with rent, or drinking water tariffs) are worth being considered. The message concluding part II seems to be that further research into the economic and financial aspects of the Modernized Mixtures especially at the local level seems to be of utmost importance.

Part III of the book is dedicated to two groups of actors at the extreme ends of sanitation chains. Farmers and domestic end-users take radically different positions at opposite ends of the waste water and human wastes chains, but in being at the periphery of the chains they also have some issues in common. Both groups are normally not included in the design- and decision-making processes about the technologies and systems to be constructed for collection, transport and treatment of the material flows which form the basis of the chains. They both feel excluded from major decisions which in principle could effect their daily practices to a considerable extent. Part III opens with two contributions focusing on the inclusion of farmers as end-users of wastes, followed by a contribution which puts the householders at the centre of analysis.

The first chapter in part III, by Jönsson, Tidåker and Stintzing explore the potential roles that farmers could play in a waste water system which is designed to economize on nutrient uses. When setting up a system for nitrogen reuse in Swedish agriculture for example, the authors argue that it is important to involve farmers with the organization of the collection, transport and spreading of the nitrogen flows. These systems should be designed in such a way that they have added value for farmers and fit into their business cycle and their main values and interests. Seven cases of urine separation are used to determine the best possible roles to be assigned to farmers in the process. The results indicate that a redefinition of the system boundaries of sanitation is hard to accomplish both for technical (hygiene, food-safety) and for social (food production and excreta do not go together) reasons.

Grendelman and Huibers in their chapter on the role of farmers as end-users of sanitation flows focus on the situation in LDCs in particular. Also in their case studies they show that a redefinition of the system boundaries in sanitation has considerable consequences for both existing and newly entering stakeholders. The design and organization of the collection, transport and treatment technologies- and systems should all be re-considered from the perspective of the agrarian end-users in peri-urban agricultural areas. For example when used in agriculture, waste-water flows should contain nutrients in certain concentrations, which makes the removal of N or P in waste water treatment not a rational thing to do in most cases. When reused in agriculture, what counts as a good quality waste water flow should be defined in close cooperation with the farmers as end-users. While some nutrients should stay in, other elements (hormones, parasites or heavy metals) should be removed in order to guarantee food safety. The general solution the authors offer is a form of 'agricultural waste water management' in which all stakeholders have a say and which takes into account all relevant factors in an integrated way.

Part III concludes with a chapter by Hegger and Van Vliet which takes a closer look at the role of domestic end-users of sanitation technologies. The concept of end-user seems awkward since strictly speaking domestic agents are the primary producers of waste-flows. They are at the origin of the flows and their behaviors and domestic practices of showering, bathing, dish-washing, cleansing and toilet-use determine both the quantity and the quality of the domestic waste-water and solid wastes flows in a direct way. The term end-user then refers to the fact that they are receivers of sanitation services at the downstream end of the sanitation chain. Similar to the farmers as discussed above, they do not take part in the major decision-making processes in the sanitation sectors. They most of the times have no say in the kinds of piped networks, treatment plants or collection trucks and bins to be selected for the separating, collecting, transporting, treating and (re)using of the waste-flows. These forms of exclusion (end-users not being consulted, informed or committed) concerning the design, construction and maintenance of the sanitation systems result in sub-optimal performances of these systems, so Hegger and Van Vliet argue. Using the Modernized Mixtures approach to analyze a number of innovative pilot projects in the Netherlands in particular, they make a strong case for including five strategic variables in the analysis and design of sanitation systems which all refer to a more active role for end-users. Domestic end-users can turn out to be important co-producers of change in sanitation if they would be taken seriously as system users, and be invited to rethink sanitation practices and to redefine their relationships to both 'neighbors' and 'nature' in experiments with new sanitation systems.

14.3 The Future Agenda of Research and Policy Making on Sanitation Systems and their End-Users

If there is one thing this book has shown, it is that the scope of social perspectives on the sanitation challenges that the world is facing today is – necessarily – wide. Much of this book dealt with socio-technical, reflexive modern ideas on how to design, organize, and implement new sanitation systems in both the North and the South. With bringing together this collection of perspectives and ideas we hope to have made clear that providing sustainable sanitation is an endeavor that is as much of a social, cultural and political character as it is of a technical one. Looking back to the various case studies of sanitation projects in both the developed world as in developing countries that have been presented in this volume, it strikes us that most of the reasons for success or failure can be found in the social, political or cultural domains, rather than in the technical domains. Although this volume presents a wide scope of social perspectives on the sanitation challenge, we should also realize that this is only a good start of thinking along these lines. If we take the claim seriously that social, political and cultural issues do matter in sanitation, further efforts in these respects towards research and development of sanitation projects are needed. To mention a few: Multi Criteria Decision Approaches need to be refined and adjusted towards local stakeholder needs, as such approaches are still rather expert focused. And the Modernized Mixture approach needs to be further developed with a more refined set of socio-technical variables that is adjusted – again – to various local circumstances. In general we need to bring the so-called 'software' much more on the front stage of political decision-making and sanitation design along with the 'hardware' of toilet systems and infrastructures.

Lastly it is time to start experimenting with new sanitation concepts in a different way. Experimentation is needed with new organization and decision-making models around sanitation, in which technology is the dependent variable instead of the other way around. From a conventional sanitation development perspective, most ideas about experimentation take toilet systems or treatment technologies as a starting point for innovation and experimentation, which can be called an 'inside-out perspective' towards the sanitation chain. We now suggest to turn it around: to look for innovations 'from the outside-in', that is: by carefully addressing the perspectives of billions of end-users and farmers at the 'Base of the Pyramid' (Hammond et al. 2007) and then designing for them and with them the socio-technical sanitation solutions that could meet their needs. What does the sanitation challenge look like if we take seriously the needs of these billions of people for sustainable, affordable, accessible, 'flush-and-forget' and trustworthy nutrient delivering systems? If we take this question as a starting point for meeting the sanitation challenges that the world is facing, the recruitment and joint efforts of both social and technical expertise is definitely needed. With this volume we hope to have shown why taking such an integrated, interdisciplinary and North-South oriented approach is worth the effort.

References

Hammond, A., Kramer, W. J., Tran, J., Katz, R., & Walker, C. (2007). *The next 4 billion: Market size and business strategy at the base of the pyramid*. Washington, DC: World Resources Institute.

Van Zon, H. (1986). *Een zeer onfrisse geschiedenis: studies over niet – industriële vervuiling in Nederland, 1850 – 1920* (A very dirty affair: studies in non-industrial pollution in The Netherlands, 1850 – 1920). Ph.D. thesis, Groningen University, Groningen.

Van Vliet, B., Chappells, H. & Shove, E. (2005). *Infrastructures of Consumption: Environmental Innovation in the Utility Industries*. London: Earthscan.

WCED. (1987). *Our common future*. Oxford: Oxford University Press.

Index

A
Abstract systems, 32, 33, 36, 37, 45
Academics, 23, 115, 123
Acceptability, 113, 118, 120
Acceptance, 21, 22, 27, 44, 58, 99–101, 185, 194–196, 206
Acceptance of new technologies, 143
Accessibility, 16, 51, 146, 155, 218
Access improved sanitation services, 105
Access point, 37
Access to drinking water, 1, 163, 167
Access to information, 72, 120
Access to land, 21
Access to on-site and semi collective sanitation facilities, 166, 169, 170
Access to sanitation, 50, 164–166, 175
Access to treated and untreated water, 199
Access to waste water, 192
Access to water, sanitation and hygiene, 111, 112
Accra, 71
Activated sludge reactors, 129
Actor networks, 193
Addis Ababa, 71
Affluent class, 53
Africa's sanitation needs, 166
African cities, 3, 11–28, 145–147, 160, 161, 165
Agrarian end-users, 224
Agricultural experts, 183
Agricultural waste water management, 7, 224
Agriculture, 2, 51, 58, 59, 70, 81, 99, 107, 161, 173, 180, 186, 191–194, 198, 223, 224
Agriculture and nature as a treatment step, 193
Agronomic conditions, 191
Allowed fertilizer, 184
Alphalog, 108
Alternative sanitation systems, 143
Ammonia losses, 186
Anaerobic digester, 101, 138, 139, 209
Anaerobic digestion, 2, 41, 95, 139
Anaerobic sludge digestion, 98
Anaerobic treatment, 41
Animal husbandry, 107
Anthropology, 206
Application as fertilizer, 180
Appropriate technology, 116
Appropriate technology paradigm, 14
Aquaculture, 194
Aqua-privy, 101
Archetypical neighbourhoods, 131
Arnhem, 42
Arua, 51
Ascariasias infections, 192
Assessment chart, 84
Assessment criteria, 6, 115
Awareness, 22, 32, 60, 62, 93, 106, 111, 120, 142, 143, 167, 213
Awareness-raising, 54, 113, 118

B
Bamako, 172–174
Banconi, 171
Bandanai, 157
Behavioral aspects, 6
Behavior change, 111, 113, 118, 122
Beige water, 74
Best performance mix, 218
Bi-and multilateral donor agencies, 80
Big infrastructure projects, 169
Biogas plants, 139
Biogas production, 32, 213
Biological treatment, 70
Bio-sanitation centre, 101
Black water, 74, 75, 80, 84, 126, 129, 147, 164, 180–186, 210
Blockage, 97, 154
Bobo-Dioulasso, 172, 175
BOD, 149, 161

Booster toilets, 210
Bore sewer, 75, 157–159, 165
Braamwisch Hamburg, 43
Brazil, 81, 161, 198
Briquette making, 61
Brown water, 99, 127, 129, 139
Brundtland, 206, 217, 218
Bugolobi, 148
Building blocks, 19, 88, 92–93, 99, 100, 102
Bukesa, 150
Bundibugyo, 62
Burkina Faso, 106–110, 117, 120, 121, 171, 173, 175
Business-as-usual, 92
Business opportunities, 185, 186
Buying power, 219

C

Cables, 40, 219
Camdessus panel, 168, 169
Canadian International Development Agency (CIDA), 116
Capacity building, 59, 167
Capacity development, 113, 118, 120, 122
Capacity to pay, 173
Captive consumer, 4, 35
Carbon removal, 160
Catchment and satellite based sewerage, 146, 155
Catchment area, 146, 153
Central government, 51, 56, 61, 168, 171
Centralized infrastructures, 27
Centralized organization, 13
Centralized sewage, 2
Centralized sewerage network, 145
Centralized waste water treatment plants, 41
Centralized water-based system, 49, 69
Central sewage system, 2
Cereal production, 181
Change agents, 206
Charge, 51, 61, 156, 158, 166, 172, 174, 180
Chemical fertilizers, 179–185
Chernobyl, 36
Chlorine-containing detergents, 212
Choice of technical system, 115
Cholera, 50, 62
Cistern-flush toilet, 80, 95
Citizen-consumer driven experiments, 208
Citizen-consumer initiated projects, 39
Citizen-consumers, 32, 35, 39, 45, 205–208, 210
Classical engineering, 205
Cleansing material, dry, 74, 80
Climate compensation, 40
Climate proof, 217
Clogging, 97
Closed loop, 2, 39
Close the loop, 184
Closing nutrient loops, 45
Co-composting, 173
COD, 149
Co-existence of different sanitation configurations, 161
Collaboration between technical engineers and social scientists, 219
Collection of faeces, 137, 139
Collection systems for storm water, 87
Collection tanks, 137, 138, 140, 180, 182, 185, 222
Collective sanitation and wastewater treatment plants, 169
Colonialism, 25
Colonial sewer system, 3
Combined heat and power plants, 139
Combined sewer overflows (CSO), 88
Comfort, 19, 20, 22, 32, 44, 213
Common goal, 199
Common language for the technical communication about sanitation, 221
Common vision, 187
Communal (infra)structures, 27
Communication methods, 115
Community-based environmental service arrangements, 27
Community based organizations, 14, 22–26, 51, 59, 64
Community hygiene, 63
Community involvement, 54, 56, 58
Community-led total sanitation, 172
Community mobilization, 59
Community-oriented approaches, 49
Community participation, 62, 64, 114
Comparing technologies, 70
Compost, 2, 41–43, 45, 129, 173, 183
Composters, 137, 138
Composting, 2, 14, 41–45, 61, 129, 137, 139, 173, 183, 207, 213
Composting system, 139
Composting toilets, 41–45, 129, 207, 213
Concentrations of pollutants, 135
Condominial sewer, 81, 158
Connectedness with the nature, 212–213
Connections to piped water, 41
Consensus, 60
Conspicuous consumption, 4, 40
Consumer loan, 172
Consumer response, 4
Consumers, 3–5, 32–35, 37–39, 41–45, 172, 191, 192, 205–208, 210, 212

Consumer studies, 6
Consumption, 3, 4, 14, 32–35, 37–41, 44, 45, 88, 129, 134, 135, 147, 165, 192, 206, 207, 213, 219
Consumption routines, 37
Convenience, 4, 53, 115
Conventional industrial fertilizers, 173
Conventional large-scale systems, 18
Conventional planning approach, 71
Corporate social responsibility strategy, 209
Correctness with the nature, 212–213
Corruption, 27, 168
Cost-effective, 23, 152, 158
Cost recovery, 14, 55, 60, 114, 164, 169, 173–174
Cradle to grave to cradle, 74
Credit systems, 114
Criteria, 6, 16, 17, 73, 84, 88, 89, 107–120, 122, 123, 146, 150, 159, 161, 172, 197, 199, 219, 221, 225
Criteria for health, 116
Criteria put forward by local actors and institutions, 221
Cropping seasons, 192
Crop water requirements, 197, 198
Culemborg, 42, 204, 208–215
Cultivation of crops, 181
Cultural acceptability, 113, 118, 120
Cultural background, 134
Cultural beliefs, 62
Cultural beliefs and practices, 22, 62, 220
Cultural dynamics, 20, 21
Cultural mix, 53
Cultural practices, 18, 142, 198
Cultural prescriptions, 21
Cultural values, 22
Culture, 20, 53, 106, 113, 115, 199, 219, 220, 222
Cycle of disease, 106

D

Daily practice of wastewater use, 189, 223
Dakar, 172–174, 198
Danish International Development Agency (DANIDA), 173
Dar es Salaam, 22, 23, 173
Data collection techniques, 117
Debate on sustainability, 106
Decentralization policy, 24, 56
Decentralized, 3, 14, 15, 18, 19, 35, 40, 50, 51, 58, 61, 63, 70, 98, 126, 127, 129, 136, 138, 139, 142, 146, 153, 161, 171, 204, 205, 209, 218
Decentralized sanitation and reuse (DeSaR), 3, 14, 15, 18, 33, 51, 60, 209
Decentralized solutions, 15, 35, 70, 126, 204
Decision makers, 89, 163
Decision-making, 5, 15, 22, 24, 89, 100–103, 107, 117, 170, 193, 194, 196, 198, 221–224
Decision-making procedure, 102
Decision-making process, 22, 221, 223, 224
Decision-making structures, 5, 15, 158
Decision-making tools, 4–7, 107, 159
Decision support systems, 218, 221
Decreased water consumption, 213
Deep-green niche markets, 44
Degradation of existing infrastructures, 56
Degree of acceptance, 195
Degree of choice of inhabitants, 205
Degree of differentiation of wastewater flows, 205
Degree of environmental consciousness, 210
Degree of tolerance, 214
Demand and supply issues, 194
Demand-based wastewater management, 194
Demand for irrigation water, 99
Demand of the reusable products, 98
Demands for sanitation, 106
Demand side management, 34
Democratic governments, 26
Demographic, 126, 147, 150, 164, 165, 190
Demographic developments, 150
Demographic growth rate, 164
Demographic transformation, 126
Densely population settlements
Dependence on international donors, 175
Dependencies between end-users, 207, 213, 215
DeSaR. *See* Decentralized sanitation and reuse
Design, 3, 5, 7, 15, 21, 26, 28, 33, 39, 45, 54, 70–76, 81, 84, 88–90, 99, 100, 102, 106, 114, 118, 120, 122, 123, 126, 131, 137, 147–149, 153, 157, 158, 160, 161, 170–172, 189–200, 207, 210, 212, 215, 217–221, 223–225
Design principles, 35, 36
Developed world, 3–5, 190, 225
Developing countries, 3, 13–15, 69, 73, 88, 98, 149, 169, 175, 190, 193, 195, 221, 225
Developing world, 2, 4, 5, 20, 164, 218
Development of sanitary systems, 20
Development path, 3
Devolution of power, 24
Dialogue about sanitation systems, 221
Diarrhea, 50
Diarrheal disease, 62
Dichotomy, 3, 15

Differentiation in service provision, 35
Differentiation of services, 38, 40, 41
Digesters, 101, 138, 139, 209
Digestion of black water, 126
Dilution, 2, 180, 183
Dimensions and different levels of scale, 18
Directorate of Water Development (DWD), 13, 58, 60
Discharge, 14, 18, 20, 27, 42, 55, 70, 72, 88, 97, 98, 129, 146–149, 154, 160, 191, 197, 199
Disconnection of rainwater, 208
Disposal, 17, 19, 25, 56, 64, 71, 72, 74–76, 88, 106, 164, 173
Distances of sewerage lines, 190
Diversified practices of consumption, 39
Domestic and industrial flows, 193, 197
Domestic consumers, 32
Domestic consumption and utility services, 33
Domestic rationalities, 19
Domestic sewage, 147
Domestic storage facilities, 22
Domestic unwanted water, 93
Domestic water usage, 4
Dominant regimes, 18
Downstream demands, 197
Downstream environmental effects, 195
Downstream flood zones, 165
Downstream peri-urban agricultural areas, 191
Downstream pollution, 191
Downstream recipients, 190
Downstream requirements, 198
Downstream use, 198
Downstream water users, 193
Drainage, 6, 12, 21, 33, 70, 87–102, 148, 156, 158, 193
Drainage and sanitation infrastructure, 89, 90
Drainage canals, 70, 88
Drivers, 114, 184–185, 187
Dry-weather flow, 97
Dual water supply systems, 42
Durban, 80
Duty and stand-by basis, 153
Dwelling, 32, 53, 140

E

East Africa, 7, 11–28, 49, 50, 59, 100, 145–161, 173, 219, 220, 222
East African Communities Organization for the Management of Lake Victoria (ECOVIC), 59
Eastern trunk sewer, 152
Eco character, 212
Eco-efficiency, 38, 146, 208
Eco-gated communities, 212
Ecological consciousness, 134
Ecological farming, 184
Ecological smell, 43
Ecological sustainability, 17, 217, 222
Eco-neighborhood, 40, 42
Economically profitable, 184
Economic and financial issues, 73, 81
Economic and political tools, 165
Economic and technological dimension, 38
Economic criteria, 113–114
Economic desirability, 90, 91
Economic performance, 34
Economies of scales, 15, 34
Eco-sanitation, 2, 3, 5
Education, 24, 59, 60, 62, 71, 108, 112, 113, 117, 120, 121, 142, 166, 167, 186, 199, 204, 209
Educational and training programs, 112
EET-DESAR, 209
Efficient tariff system, 169
Efficient waste water treatment
Effluent quality criteria, 199
Embedding of sanitation systems in local culture and politics, 222
Empirical research, 204
Empowering community, 113
Empowerment of local government, 107
Emptying latrines and septic tanks, 61
End-user perspective, 4, 7, 203–215
End-user roles, 7
End-users, 3–7, 16, 20, 22, 32, 33, 35–38, 108, 203–215, 219–225
End-users' socio-cultural concerns, 206, 207, 215
Energy crops, 183, 185
Energy recovery, 98, 129
Enforcement of rules, 199
Engineering language, 32
Engineers Without Borders, 116
Entrepreneurs, 25, 61, 64, 158, 180, 182, 185–187
Entrepreneurs for collection, storage and spreading, 182
Environmental engineers, 1, 7, 32, 43, 193, 198–200, 209
Environmental flows approach, 18
Environmental infrastructure, 11–28, 49, 50, 164, 217, 218
Environmental management institutions, 57
Environmental pollution, 53, 56, 71, 157
Epidemics, 50

Ethekwini District, 80
EU-research project, 221
European Commission, 70
European Union, 173
EVA-Centre, 210
Eva Lanxmeer, 208, 210, 212, 213
Everyday life, 4, 32, 35
Exchange of experiences, 187, 196
Excreta, 7, 64, 71, 72, 74, 76, 101, 106, 113, 164, 179, 180, 223
Excreta and faecal sludge, 72
Excreta collection unit, 106
Execution strategy, 121
Experimentation with composting toilets, 41
Expert interviews, 147
Expert-led experiments, 208
Extra asset in quality of living, 212

F

Facilitate the communication between experts and researchers, 221
Facilitator, 88, 92
Faecal matter, 93, 101, 173
Faecal sludge, 70–72, 74, 173, 174
Faecal sludge collection, 70
Faecal sludge treatment, 173, 174
Faeces, 41, 62, 70, 74, 78–80, 99, 126, 127, 129, 137, 180, 185, 186
Faeces collection, dry, 138, 139
Faisalabad, 199
Family distribution, 131
Family size, 131, 140
Farmer's role, 179–188, 223
Farmers, 5, 7, 179–200, 223–225
Farmers as end-users of wastes, 223
Farming practices, 185, 192, 193
Farmland, 7, 180
Feasibility studies, 166, 167
Feasible sanitation systems, 84
Feed back, 37, 38, 173, 187, 190
Fertilization plan, 182, 185
Fertilizer, 173, 179–186, 192
Final plan, 118, 121
Finance repairs, 174
Finance structures, 7, 164
Financial arrangements, 16
Financial mechanisms, 72, 114, 164, 170
Financial resources, 18, 51, 61, 120, 137, 167, 169, 170, 175
Financial viability of the system, 114
Financing sanitation, 166–168, 170–174, 223
Financing water for all, 168
Finite resource, 184
Fixed collection cycle, 126, 140
Fixed connection fees, 156
Flexibility, 3, 16, 17, 20, 23, 27, 64, 81, 115, 120, 126, 149, 157, 158, 160, 161
Flexible collection cycle, 126
Flooding, 56, 101
Floor-space index, 125, 132
Flow of financial resources, 120
Flows, 2, 7, 18–20, 25, 27, 32, 34, 35, 37–45, 65, 72, 76, 81, 98, 126–131, 136, 145–161, 170, 172, 179, 184, 190, 191, 193, 195, 197, 199, 205, 206, 219, 220, 222–224
Flows density, 146, 147, 149–152
Flows of varying water quality, 197
Flowstream approach, 6, 69–85
Flowstream concept, 76, 221
Flush-and-forget, 2, 220, 225
Flushing toilets with low-quality water, 42
Flushing water, 70, 74, 129, 137
Flush toilet, 2, 4, 5, 21, 42, 55, 70, 76, 80, 88, 93, 95, 97–101, 139, 182, 187
Focus group discussions, 53, 57, 62
Fodder, 184, 185
Food, 20, 31, 32, 37, 51, 174, 180, 184, 190–193, 195, 223, 224
Food chain, 31, 37, 191, 192
Food security, 20, 190, 195
Formal and informal institutions, 20, 21
Formal regulations, 21
Fort Portal, 51
Framework of flows, networks and spaces, 161
French Ministry of Foreign Affairs, 170
Functionality of wastewater treatment processes, 212
Functionality *vs.* Social distinction, 206–207, 211–212, 215
Functional perspective, 206

G

Garbage removal, 12
GDPs per capita, 107
Gender, 54, 58, 73, 81
General Directorate of Water Resource, 108
Germany, 6, 126, 135, 139, 140, 204, 222
Ghana, 71
GIE Diabeso Saniya, 171
Gothenburg, 180
Governance, 3, 5, 6, 16, 24–28, 161, 189, 190, 194–199
Governance of resources, 195
Governance of sanitation, 3
Governance principles, 189, 195, 198, 199

Governance theories, 194
Government-managed systems, 14
Gravity flow, 80, 198
Gravity sewers, 153, 154, 158
Grey water, 40, 42, 43, 70, 74, 75, 78–80, 93, 97, 99, 101, 126, 127, 129, 140, 210
Grey water recycling, 35, 129
Grid-based systems, 15
Groups of stakeholders, 114
Gurria Task Force, 168–169
Gustavsberg toilets, 210

H

Habits, 2, 4, 22, 143, 219
Hamburg, 7, 44, 131
Hamburg-Harvestehude, 131
Hamburg-Steilshoop, 131
Hand-carried water, 97
Handicraft making from waste, 61
Hand washing campaign, 63
Hand washing facilities, 54
Hanover, 139
Hazardous substances, 194
Healthcare, 174
Health concerns, 53, 111
Health hazards, 12
Health issues, 73, 81
Heavy metals, 192, 197, 198
Helminth eggs, 192
Hierarchic organization structures, 34
High demand for safety, 146
High-flush toilets, 88
High level of poverty, 107
High value domestic or industrial use, 193
History of sanitary infrastructure, 13
Ho Chi Minh City, 6, 88, 100
Hof level, 213
Hoima, 62
Hookworm, 192
Horizontal and vertical linkages, 22
Household appliances, 88
Household-based treatment systems, 193
Household composting, 14
Household dynamics, 20
Household investments, 163
Household practices, 19
Household subsidies, 170–173, 175
Household water, 41–44, 97, 135
Household water project, 41–43
Housing density, 88, 97, 140
Housing layout, 138, 139, 142
Housing types, 131, 138, 140
Håga Ecovillage, 181–182
Human excreta, 106, 179, 180
Human health, 53, 106, 157, 173, 218
Human natural cycles of nutrient flows, 43
Human parasites, 192
Human settlement evolution, 20
Hybrid solutions, 15
Hydroconseil, 164, 170–174
Hygiene, 4, 35, 71, 167, 193
Hygiene promotion, 59, 117
Hygiene/sanitation lessons, 111
Hygienic quality of life, 218
Hygienists, 218

I

Impact on a cost comparison, 126, 143
Impact to environment. 73, 81
Implementation of new sanitation concepts, 126, 129, 131, 142, 203
Implemented socio-technical systems, 211
Improper use of the facilities, 63
Improved excreta disposal, 164
Improved on-site sanitation, 146
Improved pit latrines, 14, 55, 101
Improved sanitation, 1, 5, 6, 54, 59, 62, 105, 118, 166, 222
Incineration plants, 13
Inclusion of farmers in design, 198, 200
Inclusion of the non-expert and non-professional, 89, 221
Increase transparency and accountability, 35
Incremental and ad hoc planned urban areas, 146
In-depth analysis, 126
In-depth case studies, 164
In-depth studies102, 223
Indirect financing of sanitation and infrastructures, 223
Individual household-based solutions, 22
Industrial area, 72, 81, 121
Industrialization of American households, 34
Industrialized country model, 70
Industrial water use, 197
Influence on waste management and sanitation, 55, 64
Informal settlements, 11–13, 21, 27, 51, 62, 157
Information campaigns, 22, 38
Infra-related service, 4
Infrastructure(s)
  provision, 12, 18, 41, 45, 220
  of consumption, 32–35, 219
In-laws, 62
Innovations in sanitation, 4, 33, 62, 208
Insect and rodent infection, 63

Inside-out perspective, 225
Inspection chamber, 81
Instability, 17, 50
Institutional criteria, 113
Institutionalization, 195
Institutionalization of wastewater use, 195
Institutional learning, 5
Integrated flows perspective, 20
Integrated material chain approach, 88
Intentional environmental innovations, 212
Interaction, 18, 20, 106, 125–143, 186, 187, 193, 200, 210, 213
Interdependent actors, 194
Interest of municipality and housing corporation, 209
Intermediate level of service, 165
International donor, 18, 23, 59, 61, 64, 108, 175, 223
International financial flows, 170
International Monetary Fund (IMF), 168
International Year of Sanitation, 106
Intervention cycle, 89
Interviews, 6, 62, 107–116, 118, 120, 122, 147, 181, 183, 185, 204, 209, 210, 212–214
Interview study, 107–109, 118, 120, 122
In-use involvement, 207, 213–215
Inverted siphons, 153
Investments, 2, 14, 64, 140, 164, 167, 169–173, 175, 190, 191, 193
Invisible, 32–34, 36, 38
Involvement
  and commitment of citizen-consumers, 32
  of end-users, 16, 205, 213, 215
  of peri-urban farmers, 7, 193, 194
Irregular settlements, 20
Irrigation
  equipment, 183
  of food crops, 195
  of landscapes, 195
Irrigation water management, 194
ISSUE-2, 100
Ivory Coast, 169

K
Kabagala, 150
Kabale, 51
Kalabancoro, 117
Kampala, 6, 7, 13, 21, 51, 60–62, 88, 100, 101, 146–161, 222
Kampala Sanitation Program Feasibility Report, 158
Katanga, 101
Katwe II, 150
Kenya, 24, 147, 152, 158, 165, 169, 222
Kenya Breweries, 148
Key dimensions of sustainable development, 217
Key domestic routines, 34
Key entry points, 64
Key stakeholder, 180, 187
Kibaale, 62
Kibla, 21
Kibuli, 150
Kibuye II, 150
Kinawataka, 150, 155
Kisat, 148
Kisat sewage treatment works, 148
Kisugo, 150
Kisumu, 1, 25, 146-161, 222
Kisumu cotton mills, 148
Kisumu Molasses treatment works, 148
Kisumu Water and Sewerage Company (KIWASCO), 25, 147, 152
Kisutu Women Development Trust Fund (KIWODET), 23
Kiswa, 150
Knowledge gap, 107
Kololo, 156
Kvicksund, 183–185

L
Lake Victoria, 12, 55, 147
Lake Victoria Basin Commission (LVBC), 59
Lake Victoria Environment Management Project (LVEMP), 59
Lake Victoria Local Authorities Cooperation (LVRLAC), 59
Lake Victoria South Water Service Board (LVSWSB), 153
Land-locked countries, 107
Land requirements, 155, 160
Language of tokens, 221
'Large is robust and efficient', 204
Large-scale service provision, 23
Large technical networks, 43
Large technical sewage system, 3
Latrine, 2, 13, 14, 19, 22, 27, 53–55, 61–63, 71, 72, 75, 101, 118, 120, 121, 157, 165, 173
Laws and policies, 113
Laws in Uganda, 53
Learning networks, 113
Legal and policy criteria, 118
Legal aspects, 186
Legitimacy, 23, 27, 214
Lemmerweg-Oost, 208–210
Level of service, 165–166, 217

Levels of poverty, 113
Levels of trust, 37
Life cycle assessment (LCA), 108, 115, 116
Life cycle costs, 113
Lifestyle, 4, 35, 37–39, 42, 210
Life-world perspective, 208, 215
Limited acceptance, 185
Links between sustainable solutions and end-users' socio-technical concerns, 206
Liquid composting plan, 183
Lira, 51
Literature review, 108, 109, 122
Living machines, 207
Local actors, 6, 107, 109, 111, 117, 122, 221, 222
Local authorities, 14, 22, 25, 36, 49, 51, 57, 59, 64, 164, 167, 168, 174, 194, 223
Local conditions and demands, 24
Local Councils (LCs), 56
Local governments, 51, 60
Local infrastructure, 90, 92, 97
Locally-available funding, 175
Locally grounded criteria, 110, 123
Local perceptions, 107
Local water pumping and treatment facilities, 42
Lock-in effects, 17, 19, 220
Long term financing, 122, 163
Long term performance, 179
Low developed countries, 3, 4
Low priority, 50, 53, 61, 118, 166
Low-quality, 42
Low sanitation coverage, 50
Low-tech, 11
Lubigi, 150
Lund, 183–185, 187

M

Ma'an, 71
Macro-economic policies, 26
Maintaining food security, 195
Maintenance, 13, 14, 19, 28, 32, 43, 55, 56, 63, 106, 112, 147, 149, 153–155, 160, 164, 167, 170, 173, 174, 187, 190, 196, 199, 200, 213, 219, 224
Makarere, 100, 101
Makindiye, 150
Making the invisible visible, 33, 34
Mali, 105–107, 109, 110, 117, 165, 166, 171
Malian Municipal Mayor, 108
Management and maintenance, 43, 213
Management of risk, 98, 99
Management scale, 205
Manholes, 81, 153
Manure, 183, 184
Manyatta, 152, 157
Market-and consumer-demands, 34
Martin's Dyke, 154
Masaka, 51
Master and structure plans, 13
Material flow analysis (MFA), 74, 76
Material flows, 2, 18, 32, 38, 40, 193, 219, 223
Material infrastructures, 219
Matrix of products, 74
Mayuge Town Council, 51
Mbale, 51
Mbuya, 150
MCDA tool, 88, 90
Meaning categorization approach, 108
Mechanized aerobic treatment systems, 145
Medium-large-scale public sewerage, 161, 165
Medium-term expenditure frameworks, 167
Mengo, 150
Methane, 2, 129
Metropolitan area, 156, 192, 193
Mexico, 161, 168, 198
Mexico City, 198
Miasmas, 35
Miasmatic theory, 35
Micro-credit, 172, 175, 223
Micro-finance, 170–172
Micro-flush toilets, 76
Middle East, 191
Milimani, 152, 156, 158
Millennium Development Goal (MDGs), 1, 4, 12, 59, 64, 69, 105, 106, 117, 163–165, 167, 175, 218
Mini-sewer users, 172
Minister of Finance, 168
Ministries, 59, 166, 204, 209
Misuse of toilet facilities, 56
Mixing of waste water flows, 126
Mobility concepts, 210
Modernization path to be chosen, 218
Modernized mixture, 3, 5–7, 12, 15–28, 51, 64, 146, 157, 161, 204–208, 215, 218–225
Modern technology, 5, 205
Modular approach, 16, 28
Multi-criteria analysis, 6
Multi-criteria decision analysis (MCDAs), 89, 108, 115, 221
Multi-level governance, 6
Multiple pathways, 27
Multiple scales, 16, 18
Municipal authorities, 11, 55
Municipality owned companies, 34
Municipal level, 117, 121
Municipal wastewater treatment plant, 183
Muslim, 21, 22, 220

Mutungo, 150
Mwanza, 21

N
Naalya, 148
Naguro, 150
Nairobi road, 154
Najjanakumbi, 150
Nakawa, 150
Nakivubo, 150, 154, 160, 161
Nalukolongo, 150
Namirembe, 150
National Environment Management Authority (NEMA), 54, 56, 58, 149
National Office of Water and Sanitation, 108
National Water and Sewerage Corporation (NWSC), 13, 56, 58, 148
National Water Policy (NWP), 60
National Water Sewerage Corporation (NWSC), 13, 56, 58, 60, 148
Natural ditches, 88
Needs and demands of end-users, 207
Needs for laws, policies and institutional frameworks, 113
Negowat, 198
Neo-development state, 26
Netherlands, 33, 41–44, 126, 203, 204, 210–212, 224
Network approach, 25–27
Network-bound systems, 215
Network governance approach, 26, 27
Networks, 7, 18, 19, 25–28, 33–35, 39–41, 43, 55, 59, 70, 72, 81, 113, 121, 139, 145–161, 165, 174, 190, 193, 194, 215, 219, 222, 224
New concepts for sustainable socio-technical systems, 205
New sanitation regime, 5, 217
Niche management, 5, 205
Niche markets, 35, 44
Niche projects, 215
Niches, 5, 18, 21, 35, 44, 205, 206, 215, 217
Nigeria, 169
Night soil, 154
Nile Basin Initiative (NBI), 59
Nitrogen (N), 33, 74, 75, 128, 149, 157, 161, 180–184, 191, 192, 223, 224
Nobler image, 168
Non-Olet, 41–43
Non-paid professionals, 221
Normative claims about the future and-users, 214
Norms and values of end-users, 22
Norrköping, 182, 184, 185, 187
Ntinda, 148, 150, 151
Nuisance from the noise, 209
Nuisance of flies, 43
Nuisance of smell and sight, 43
Nutrient content of the urine, 181
Nutrient flow, 43, 179, 784
Nutrient load, 75, 76
Nutrient reuse, 32
Nutrient-rich flows, 127
Nutrients, 2, 7, 32, 43, 45, 70, 74–76, 78, 98, 99, 126–129, 160, 179–185, 191–194, 197–199, 206, 213, 217, 222–225
Nutrient uses, 223
Nyalenda, 148, 152, 154, 157, 158

O
Oats, 181
Objective of the system, 180
Objectives, 12, 16, 59, 60, 70, 73, 88–93, 97, 99,–102, 107, 108, 126, 170–171, 180, 195
Obunga, 157
OECD countries, 13, 19, 34, 205, 220
Office National de l'Eau et de l'Assainissement of Burkina Faso, 173
Official development aid, 23
Official Development Assistance, 168
Off-site, 88, 93, 95, 97, 98, 101
Oil crises, 34
One-approach-fits-all, 123
One fit for all, 218
On-site, 2, 22, 35, 36, 41, 55, 71, 72, 75, 88, 93, 97–99, 101, 108, 120, 121, 129, 146–148, 157–159, 165, 166, 169, 170, 172, 175, 182, 223
On-site collection, Storage and Treatment, 75
On-site composting, 2
On-site pre-treatment, 88, 99
On-site sanitation facilities, 72, 170, 172, 175
On-site sanitation systems, 55, 147, 148, 175, 182
On-site storage, 22
On-site systems, 2, 158, 159, 182
Open-defecation free community, 172
Operational costs, 14, 140, 143, 173
Operation and maintenance costs, 149, 155, 160, 170, 173, 214
Opportunities, 6, 7, 12, 23, 33, 40, 50, 53, 61, 64, 101, 121, 185, 190, 192, 198
Organizational set-up, 16
Ouagadougou, 71, 81, 109, 120, 172, 173, 175
Outbreak of diseases, 57
Overcrowded areas, 165

P
Package deal, 214
Pakistan, 199
Paradigmatic shift, 35
Paradigm shift, 33, 40, 122, 220
Parasites, 192, 224
Paris, 180
Participation of citizen-consumers, 205
Participation of stakeholders, 27, 102, 120
Participation of users in decision-making, 196
Participative planning, 118
Participatory design, 88
Participatory design and operation, 195
Participatory methods, 114, 116, 117, 121
Participatory observation, 204, 211
Participatory processes, 196, 198
Particular demands, 21
Partnership, 54, 57, 61, 147
Path dependencies, 3
Pathogen and nutrient removals, 160
Pathogenic microorganisms, 192
Pathogens, 32, 36, 45, 75, 191, 194
Pathogen spread, 74
Pathogens removal, 32, 45
Pathogen transmission, 63
Payment facilities, 172
Payment system, 5, 15, 158, 204
Peak flows, 136
Penalize, 173
Perceptions of sustainability, 105–123
Performance matrix, 6
Peri-urban areas, 110, 146, 165, 170, 190–192, 194–197
Peri-urban farmers, 7, 193–195
Peri-urban poor, 6
Perspective of developing countries, 169
Perspective of end-users, 108
Pharmaceuticals, 134
Phosphate, 2
Phosphorus, 76, 182–184, 192
Physical and policy requirements, 99
Physical characteristics, 6, 131
Physical conditions, 90, 92, 93, 97, 98, 117, 159
Pilot phase, 121
Pilot project, 4, 5, 7, 39, 121, 126, 175, 203, 204, 206, 208–211, 213–215, 217, 224
Pilot project approach, 179
Piped water, 12, 13, 41, 97
Pipe requirements, 140, 142
Pit latrines, 2, 14, 19, 27, 55, 61, 63, 101, 118, 157, 165
Plan Helvetas, 108
Planned growth in urban services, 51
Plant fertilizer, 126, 127
Plant nutrients, 129, 180–182, 184
Playground for a transition in wastewater systems, 205
Plot sizes, 142
Policy gaps, 25
Policy statements, 185
Policy tool, 84
Political and public concern, 166
Political feasibility of the goals, 218
Political instability, 50
Politicisation, 27
Pollution of the surface water, 190
Poor communities, 51, 56, 60, 62
Popular version, 57
Population equivalent, 147
Population growth, 19, 50, 62, 105, 192
Population influx, 64
Potassium, 192
Potential driving forces for transition towards sustainability, 206
Pour-flush latrine, 72, 75
Poverty alleviation, 59, 195
Poverty Eradication Action Plan, 59
Poverty reduction Strategy Papers, 169
Power of culture, 220
Power of end-users, 207–208
Power relations, 196, 221
Power vis-α-vis systems of provision, 214, 215
Practical feasibility, 93
Practitioners, 122, 123, 126, 219, 221
Pre-condition for success, 223
Pregnant women, 62
Preliminary agreements, 121
Prescriptive engineers, 200
Presence of pathogens, 191
Pressure flow, 80
Pre-treat, 72, 88, 99
Pre-treatment plants, 148
Preventing diseases, 218
Primary collection, 19
Primary objectives, 90
Principle actors, 117
Private and quasi-public institutions, 161
Private investment, 170–173, 175
Private participation, 3
Privatization, 14, 34, 42
Process criteria, 113–115, 118, 120
Product-Process, 76
Product separation, 74, 80
Product-specific method, 76
Product-specific process, 76
Promoting social interaction, 210
Promotional project, 121
Promotion of improved sanitation, 118

Proper use of the facilities, 142
Pro-poor sanitation programs, 64
Pro-sanitation policies, 113
Protect new technologies from the mainstream market, 206
Protects public and environmental health, 106
Protozoa, 162
PROVIDE, 11, 12, 19, 84, 100
Provide and instruct, 45
Provider-client relationship, 24
Provider-related experts, 102
Providers, 2–4, 32, 34, 37–39, 41, 42, 45, 112, 211, 212, 214, 219
PR strategy, 209
Ps-Eau, 164, 170–174
Public driven sewerage developments, 150
Public health, 20, 34, 55, 57, 90, 105, 146, 157, 159, 161, 191–193, 195, 212
Public health threats, 157
Publicly-managed, 15
Public perceptions, 194
Public-private partnerships, 24, 60
Public sector management, 24

Q

Quality level, 190, 192
Quality of life for citizen-consumers, 39
Quality of social institutions, 25
Quality of their consumption practices, 206

R

Radical innovation, 18, 208
Ragn-Sells, 183
Rainfall intensities, 97, 98
Rainfall regime, 97, 99
Rainwater, 40, 70, 97, 127, 199, 208, 210
Raised pit latrine, 101
Ranking chart, 84, 87
Rapid demographic changes,190
Rational resource use, 33, 42–44
R&D project, 181
Receivers of sanitation services, 224
Reconfiguration of sanitation flows, 157
Recycled faecal matter, 173
Recycled product, 7, 180
Recycling, 28, 61, 126, 127, 129, 173, 180, 182–187
Recycling sanitation systems, 7, 180
Reduce the application of chemicalfertilizer, 182
Reed bed filters, 42, 43, 207, 212–214
Reed odourless earth closet, 157
Reflexive engineers, 200
Reflexive modernization, 31, 218
Refurbishment of existing buildings, 126
Regional and national grids, 32
Regional Centre for Low-Cost Water and Sanitation, 108
Reimbursement facilities, 170
Re-introducing the senses, 220
Religious prescriptions, 220
Re-localization, 45
Replacement of chemical fertilizers, 184
Re-sensitization, 36, 40
Re-sensitization at different levels, 45
Resilient, 16
Resource allocation, 53
Resource efficiency, 143
Resource-efficient wastewater management, 127
Resource oriented sanitation, 2
Responsibilities for sanitation, 166
Responsibility, 14, 26, 55, 56, 174, 182, 183, 207, 209
Restrictions of each stakeholder, 187
Retrofitting, 126, 139, 141, 143
Reuse in agriculture, 70, 81, 99
Reuse-oriented systems, 88
Reuse-oriented toilet types, 92
Reuse possibilities, 92
Reuse potential, 76
Reversed wastewater chain, 7
Risk behavior, 212
Risk for wastewater management systems, 212
Risk politics, 32
Risk Society, 36
Role of consumers, 35
Role of domestic end-users, 203, 224
Role of domestic end-users of sanitation technologies, 224
Role of experts, 93
Role solar panels, 42
Rufisque, 172, 174
Rule-making and steering mechanisms, 194
Rural-urban migration, 51

S

Safeguard and promote public health, 57
Safeguarding hygiene, 193
Salinity, 192, 197
Sanitary configurations, 146, 147
Sanitary flows, 41, 146, 148–152, 157
Sanitary landfills, 13
Sanitary space characteristics, 147
Sanitation, 1–4, 11, 31–46, 49, 69, 87–102, 105–123, 126, 145, 164, 186, 190, 203, 218

Sanitation behaviour, 62
Sanitation chain, 2, 7, 16, 88, 100, 223, 225
Sanitation Challenges, 4, 5, 14, 218, 224, 225
Sanitation crisis, 164–166
Sanitation flow chain, 2
Sanitation infrastructures, lack of, 220
Sanitation process, 171
Sanitation projects in developing contexts, 71
Sanitation promotion, 118, 167
Sanitation provision, 41, 50, 52, 122, 157, 221, 222
Sanitation regulations, 147
Sanitation sector, 57, 59–62, 108, 117, 122, 123, 152, 163, 168, 170, 174, 224
Sanitation service provision, 3, 6, 45, 51
Sanitation surcharge, 175
Sanitation surtax, 120
Sanitation system, 3, 5–7, 17, 20, 21, 32, 35, 36, 40, 43, 44, 51, 55, 69–85, 87–92, 97, 98, 105, 106, 111–116, 122, 123, 126, 131, 139, 142, 143, 146–148, 155, 156, 159, 165–167, 175, 179–188, 205, 207, 212, 215, 217, 219, 221, 222, 224–225
Sanitation technology, 3, 5, 161
Sanitize the collected products, 185
Sanplats, 55
Satellite sewerage, 158, 161
Satellite system, 146
Scale of service, 41
Scepticism, 43, 180, 186
Science and Technology Studies, 2–3, 6
Screening, 88, 90, 93–100, 102
Secondary collection, 19
Sedimentation technologies, 70
Self-organization, 22
Self-sufficient system, 127, 129
Semi-arid savanna, 107
Semi-centralized systems, 190
Semi-collective sanitation programmes, 170, 174
Senegal, 165, 171, 175, 198
Sense, 6, 31–46, 108, 109, 113, 196, 212, 220
Sensitization, 39, 41–46
Sensitization of practices, 43
Sensitizing infrastructures, 36
Sensitizing material flows, 40
Sensitizing sanitation, 41, 45
Sensory experiences, 32–36, 38, 41
Separate collection of the black wastewater, 209
Separate piped water system, 41
Septic tank, 19, 20, 55, 61, 63, 75, 76, 88, 92, 99, 101, 150, 154, 157–159, 161, 165
Sets of environmental problems occurring at once, 190
Sewage system, 1–3, 5, 13, 70, 147, 154, 155, 187, 208, 210
Sewage treatment points, 55
Sewerage modernization, 145
Sewerage network, 72, 81, 121, 145, 150, 152, 153
Sewerage treatment works, 150, 158
Sewer systems, 3, 35, 41, 80, 88, 97, 99, 154, 207
Sewer treatment, 2, 99
Shallow sewer, 72, 158, 159
Simple modernization model, 218
Simplified sewers, 81
Siphons, 153–155, 157
Site-specific objectives, 92, 99, 102
Sitting-down toilets, 22
Situational analysis, 121
Size of households, 131
Skills, 60, 92, 112, 168, 196, 199, 207
Slabs, 118
Slum dwelling poor, 53
Slums, 5, 51, 53, 57, 62, 165, 218, 220
Slum settlements, 146
Slum upgrading programmes, 158
Slurry tanks, 186
'Small is beautiful', 204, 205
Small-scale, 14, 15, 60, 61, 150, 158, 165, 174, 206
Small-scale enterprises, 60, 61
Small-scale experiments, 206
Small-scale satellite systems, 150
Small technology, 15
Smell, 32, 35, 40, 43–45, 209, 214
Sneek, 204, 208–215
Soakage pit, 88, 101
Social and cultural norms (relevance of), 36, 222
Social and institutional barriers against, 206
Social, cultural and gender issues, 73, 81
Social-cultural norms and practices, 22
Social development tools, 120
Social dimension of sustainable development, 218
Social dimensions of the sanitation challenge, 219
Social-institutional aspects, 111
Social learning, 196
Socially acceptable, 106
Social marketing techniques, 60
Social practices, 3, 18, 19, 22, 27
Social sciences, 3
Social scientific perspective, 2, 45

Social status, 53, 62
Social trust, 13
Socio-cultural concerns of end-users, 204, 207, 211
Socio-cultural meaning, 2, 206
Socio-cultural symbols, 206
Socio-economic management arrangements, 12
Sociology of consumption, 206
Socio-technical innovation, 204, 208
Socio-technical orientation, 147
Software, 108, 225
Soil compaction, 181, 183
Soil improvers, 173
Soil permeability, 159
Soil salinity, 192
Soil structure, 192
Solar panel systems, 33
Solids concentration, 154
Solid waste, 11–16, 18–20, 22, 23, 25, 27, 32, 53, 55–57, 61, 62, 74, 100, 224
Solid waste infrastructures, 14
Solid waste management, 12, 18, 53, 56, 57, 100
Soroti, 51
Source-oriented systems, 93, 99
Source separated product, 186
Source separating sanitation systems, 7, 142, 222
Source-separating sanitation technologies, 125–143
South Africa, 80, 169
Space, 7, 40, 42, 84, 97, 100, 102, 131, 132, 137, 138, 140, 142, 145–161, 222, 223
Space demands, 146, 155–157
Spatial plan, 13, 42, 204, 209
São Paulo, 198
Stabilization ponds, 72, 81, 148, 149, 160
Stage of decision-making and realization, 101
Stakeholder dialogues, 6
Stakeholder roles, 89–90, 180
Stakeholders, 16, 27, 56, 57, 64, 88–90, 92, 100, 102, 108, 113–115, 117, 120–123, 166, 168, 171, 180, 183–185, 187, 188, 193–196, 199, 200, 221, 223, 224
Stakeholder workshop, 92, 99, 102, 118, 121, 221
Stand-alone document, 84
Standardized format, 73–74
Standards and regulations, 56, 161
Standards of cleanliness, convenience, hygiene, 4
State-owned utility companies, 34
State regulator framework, 21
Stockholm Vatten company, 181
Stormwater, 74, 93, 98, 99, 101
Stormwater flow, 97
Stormwater runoff, 93, 98
Strategic city level, 146, 159
Strategic managers, 206
Strategic niche management, 205, 215
Strategic plan, 71, 121
Strategic plan for sanitation, 71, 109, 120
Strategic sanitation plan, 71, 109, 172, 175
Stream separation at source, 92, 93
Strength-weakness analysis, 100, 101
Structural Adjustment Programs (SAP), 26
Sub-Saharan Africa, 7, 69, 163–167, 169, 175, 223
Subsidies, 72, 114, 120, 171–173, 175
Successful sanitation projects, 85
Successful transition, 143
Success rates of water and sanitation projects, 106
Suitability assessment, 138
Sum of flowstreams, 76
Sunk costs, 3, 4
Supply chain, 114, 172, 173, 193, 199
Supply driven, 34, 71
Supply-oriented, 34
Surrogate indicators, 34
Sustainability criteria, 107, 109, 111, 116, 118, 122
Sustainability of the sanitation system, 179–188
Sustainable character of the neighbourhood, 210
Sustainable energy, 210
Sustainable financing mechanisms, 116, 122
Sustainable sanitation, 4, 12, 25, 43, 73, 106–108, 110–112, 115, 117, 122, 161, 222, 224
Sustainable Sanitation Alliance (SuSanA), 106
Sustainable urban development, 15
Sustainable water and waste water management, 203
Sweden, 126, 181, 184, 204
Switzerland, 126
Symbolic dimension, 38
Symbolic level, 42
Symbolic media, 37
Symbolic representations, 43
System and/or planning concepts, 70
System concept, 78
System design, 71, 73–74, 84, 122, 123
System rationalities, 19
Systems of provision, 3, 33, 35, 207–208, 214, 215
System transformation, 5

T
Taboos, 62
Take over responsibilities, 213
Tanum, 182–183, 185–187
Tanzania, 12, 21, 22, 24
Taxes, 61
Technical baseline studies, 122
Technical characteristics, 73, 81
Technical criteria, 115, 118, 122, 197
Technical design, 71, 118, 120, 137, 193
Technical dimensions, 219, 221
Technical functionality, 91, 97
Technical parameters, 193
Technical solutions, 193, 222
Technical system design, 2, 89, 115, 122
Technical systems, 2, 89, 115, 122
Technological and social dimensions, 27
Technological scale, 3, 205
Technology assessments, 92, 109, 110, 115, 116
Technology-specific, 90–92, 99, 100
Tenant-owner association, 182
Thematic maps, 117
Theoretical and policy perspective, 19
Third world cities, 5
Threshold levels, 149, 152, 156, 158, 159
Time of fertilization, 185
Time-space telescoping, 190
Timing and volumes of energy and water consumption, 34
Toilet
  absence of, 100
  composting, 41–45, 129, 207, 213
  dry, 2, 97, 101, 220
  paper, 80
  sitting-down, 22
  stand-alone, 41
  urine diversion, 2, 142, 180, 187
  users, 7, 43, 179
  waste water, 181
Top-down innovation can be combined with an end-user perspective, 210
Topography, 159
Total system performance, 180
Toxins, 197
Traditional pit latrines, 27, 55
Trajectories, 18, 204
Transfer of responsibilities, 168
Transition, 1, 4, 27, 45, 87, 143, 157, 159, 205, 206
Transition experiments, 215
Transition management, 5
Transport, 2, 4, 25, 41, 55, 72, 73, 75, 78, 80, 88, 92, 93, 96–98, 106, 140, 145, 166, 167, 173, 180, 181, 183, 185, 186, 190, 210, 220, 223, 224
Transport media, 80
Treatment capacity, 190
Treatment efficiency, 193
Treatment offsite, 75, 80
Treatment process, 106, 160, 212
Trunk sewer, 150, 152, 153
Trust of residents, 213
Trust of the water users, 195
TSS, 149
Tunisia, 175, 199
Twin alternating vault toilet, 101

U
UASB with trickling filters, 161
Uganda, 6, 7, 12, 13, 21, 24, 49–64, 87, 88, 100, 146, 165, 171, 220, 222
Uganda Water and Sanitation NGO Network, 59
Umbrella organization, 59
Unacceptable environmental damage, 34
Unauthorized dumping areas, 173
Uncontrolled discharge, 191
Uncontrolled dumping, 62
Unemployment rate, 131, 142
UNICEF, 164, 172
Unification, 87, 135
Uniform consumption practices, 39
Uniform infrastructure service provision, 34, 35
Uniform sewerage charges, 158
UNISE, 148
United Nations Development Program (UNDP), 116
United States Peace Corps, 116
Unplanned high-density communities, 53
Unplanned settlements, 218
Unsustainable land-use, 191
Untreated or partially treated wastewater, 189, 191
Upgrading existing inadequate infrastructure, 92
Upper Kitante West sewer, 153
Uppsala, 181
Upstream and downstream actors, 18
Urban agriculture, 51, 161
Urban-application context of the method, 93
Urban authorities, 13, 50, 53, 61
Urban centres, 32, 49–55, 57
Urban environmental infrastructures and services, 12, 18, 26, 27
Urban environmental services, 26, 50, 53–56
Urban form, 6, 125–143

Urban infrastructure provision, 18
Urban infrastructures, 11, 35
Urbanization, 11, 13, 49–53, 62, 87, 88, 105, 146, 165
Urban planning and management, 51
Urban planning and modernization, 146
Urban poor, 3, 15, 20, 25, 49–64, 72, 169, 218
Urban sanitation models, 218
Urban sanitation upgrading, 100
Urban slums, 5, 51, 53, 55, 218, 220
Urine, 2, 41, 45, 74, 78, 80–83, 93, 126–129, 137, 138, 180, 181, 184–186
Urine and faeces separation, 70, 78
Urine-diversion, 2
Urine diversion or vacuum toilets, 2, 98, 99, 180
Urine diversion toilet, 2, 142, 187
Urine-diverting dehydrating toilets, 93, 95
Urine diverting dry toilets (UDDTs), 80, 97–99, 101
Urine-diverting flush toilet, 95, 98, 99
Urine separation, 125, 126, 128, 137–142, 181–183, 222, 223
Usage facilities, 113
Usage of nutrients, 191
Use-based design, 199
Use of domestic wastewater, 191
Use of open space, 142
Use of polluted water, 189
Use of sanitary technologies, 22
Use of sewage sludge in agriculture, 180
Use of sludge in agriculture, 173
Use of the effluent, 193
Use of untreated or partly treated wastewater, 191
User acceptance, 22, 99–101
User choice, 114, 115, 122
User interface, 75, 76, 78, 80, 106
User-related objectives, 101
Users and re-users of products, 88
Using source separated human urine, 181
Utrecht, 41–43, 45

V

Vacuum sewers, 138
Vacuum stations, 138, 139
Vacuum technology, 80
Vacuum toilet, 2, 41, 43, 44, 93, 95, 98, 99, 139, 180, 183, 207, 209
Vacuum trucks, 137, 140, 173
Ventilated improved pit (VIP), 55, 72, 101, 120
Vertical integration, 199
Vietnam, 6, 88, 97, 100
Visibility, enhanced, 37
Visibility of reed bed filters, 42
Visible water drainage, 33
Visualization of the material flows, 40
Voluntary contributions, 22

W

Wabigalo, 150
Wageningen, 5, 41, 43
Wandegeya, 150
Waste collection, 20, 23, 34, 61
Waste separation, 22
Waste stabilization ponds (WSP), 148, 149, 160
Waste water, 2, 3, 32, 35, 40, 45, 70, 75, 80, 81, 87, 88, 98, 99, 126, 148, 153, 164, 173, 174, 180, 181, 186, 189–200, 219, 223, 224
Waste water chain, 7, 18, 19, 120, 199
Waste water discharge quality, 146
Waste water flows estimated for sewerage, 147
Waste water generation, 131, 133, 137–138, 142, 161
Waste water irrigation, 7, 192
Waste water management, 32, 126, 127, 189, 190, 193–195, 203–205, 207, 210, 212, 224
Waste water management teams, 196, 198, 200
Waste water reuse, 5
Waste water treatment facility, 70, 72
Waste water treatment plant (WWTP), 13, 41, 70, 72, 81, 93, 95, 98, 101, 169, 183, 199
Waste water use, 148, 189–200
Waste water volume, 135, 191
WaterAid, 109, 117–120, 122, 172
Water and sanitation sector working group (WSSWG), 59
Water based conventional sewer system, 69
Water-based infrastructure regime, 4
Waterborne and water related diseases, 50
Waterborne diseases, 192
Water City, 209
Water closet, 2
Water flush toilet, 2, 70
Water provision, 12, 20, 51
Water recycling, 35, 126, 127, 129, 180
Water-savings, 32, 38, 43, 45, 88, 208
Watersheds, 189
Water supply, 1–3, 13, 35, 42, 44, 59, 60, 88, 90, 92, 116, 168, 190, 192
Water works, 2, 34
Weakness in communication, 120

Wellness, 38, 40, 44
Well-planned residential areas, 53
Western countries, 12
Western Europe, 6, 7
WHO-UNICEF Joint Monitoring Programme, 164
Willingness to pay, 92, 121, 172
Win-win situations, 204
World Bank water and sanitation program, 115–116
World Water Forum, 168

Y

Yields, 108, 191, 192, 198
Yuck factor, 36, 43

Lightning Source UK Ltd.
Milton Keynes UK
UKOW021128180212

187416UK00005B/26/P